The Mexican Transition Zone

Juan J. Morrone

The Mexican Transition Zone

A Natural Biogeographic Laboratory to Study Biotic Assembly

Springer

Juan J. Morrone
Museo de Zoología 'Alfonso L. Herrera'
Facultad de Ciencias, UNAM
Mexico City, Mexico

ISBN 978-3-030-47916-9 ISBN 978-3-030-47917-6 (eBook)
https://doi.org/10.1007/978-3-030-47917-6

This Springer imprint is published by the registered company Springer Nature Switzerland AG.
The registered company address is: Gewerbestrasse 11, 6330 Cham, Switzerland

To Gonzalo Halffter,
who taught me that biogeography is much more diverse and complex than I had ever imagined.

There are more things in heaven and earth,
Horatio, than are dreamt of in our philosophy.
Hamlet, act I, scene V

Do I contradict myself?
Very well then I contradict myself.
(I am large, I contain multitudes.)
Walt Whitman (1892), Song of myself

Preface

I met Gonzalo Halffter twenty-one years ago. I have been invited to give a lecture on biogeography at the Instituto de Ecología, Xalapa, Veracruz. I was young (well, I thought I was young), Léon Croizat was my personal hero, panbiogeography and cladistic biogeography were the only approaches I was applying as a practicing biogeographer, and I was trying to give a good impression to my audience. After the lecture, Gonzalo asked me bluntly why I and my colleagues at UNAM considered that vicariance was the only relevant biogeographic process, dismissing dispersal at all. I felt a little uneasy, but I answered him trying to be as clear and polite as possible. After returning to Mexico City, I realized that I knew very little about Halffter's biogeographic contributions, so I decided to begin studying them. While reading them I discovered that there were other "dispersalists," like Osvaldo Reig and Jay Savage, who held ideas similar to Halffter's that both dispersal and vicariance were relevant biogeographic processes. This epiphany was surprising: these biogeographers were not the extreme dispersalists (*à la* Matthew) that I had imagined, but reasonable empirical biogeographers trying to develop an integrative approach to evolutionary biogeography.

During the following years, I had several opportunities to enjoy Gonzalo Halffter's conversation and profound knowledge. We discussed several issues, not always agreeing. My clear distinction between dispersalists and vicariance biogeographers faded away. (Conversations with Pedro Reyes Castillo and Mario Zunino were also very helpful in this respect.) Ten years ago, I developed the conviction that evolutionary biogeography was more complex than I had previously imagined, and I incorporated the dispersal–vicariance model, transition zones, and cenocrons to my perspective of biogeography. This book represents both an analysis of the Mexican Transition Zone and an empirical application of my evolutionary biogeographic perspective.

In the first chapter, I provide a general characterization of biogeographic transition zones and how they are analyzed by both the ecological and evolutionary perspectives. Several concepts are discussed and the main biogeographic transition zones of the world are briefly introduced.

In the second chapter, I present a general introduction to evolutionary biogeography, where different methods are used to answer different questions, which are considered as successive steps of an integrative analysis. I detail these steps and refer briefly to some of the methods that may be applied to answer particular biogeographic questions. I also discuss how different methods are integrated within an integrative framework, which is particularly appropriate for analyzing transition zones.

The third chapter represents a historical perspective of the Mexican Transition Zone. I refer specially to Halffter's conttributions, in a historical sequence. I analyze the development of his theory and distributional patterns recognized by him, discussing how they are considered to represent cenocrons. I refer also to other authors who have analyzed the Mexican Transition Zone, undertaking dispersal, track, cladistic, endemicity, and phylogeographic analyses.

In the fourth chapter, I analyze the biogeographic regionalization of the Mexican Transition Zone, characterizing its biogeographic provinces: Sierra Madre Occidental, Sierra Madre Oriental, Transmexican Volcanic Belt, Sierra Madre del Sur, and Chiapas Highlands. I discuss their circumscription, endemic species, biotic relationships, and vegetation. I also deal briefly with the districts that have been recognized within these provinces.

The fifth chapter deals with the biotic assembly of the Mexican Transition Zone. I characterize the original Paleoamerican biota and the four cenocrons that assembled successively to it, namely, the Mexican Plateau, Mountain Mesoamerican, Nearctic, and Typical Neotropical cenocrons. I also analyze the biotic assembly of the cenocrons, from the Cretaceous to the Holocene, as well as the Paleogene, Neogene, and Quaternary horobiotas that can be recognized in the Mexican Transition Zone.

In the last chapter, I discuss some general perspectives, especially referred to transition zones and to evolutionary, ecological, and integrative biogeography. I try to analyze how integration between the historical and ecological perspectives can be undertaken in future studies.

During the years I have benefited from my interaction with several friends and colleagues, especially important for my understanding of the Mexican Transition Zone has been Gonzalo Halffter, who has generously shared his ideas with me. Pedro Reyes Castillo and Mario Zunino were also instrumental in discussing biogeographic issues. Also important have been Roxana Acosta Gutiérrez, Manuel Barrios Izás, Enio Cano, Tania Escalante, David Espinosa Organista, Ignacio Ferro, Oscar Flores Villela, Livia León Paniagua, Jorge Llorente Bousquets, Juan Márquez-Luna, Miguel Ángel Morón, Adolfo Navarro Sigüenza, Gerardo Rodríguez-Tapia, and Margarita Santiago-Alvarado. I thank them for their patience and collaboration. Federico Escobar, Juan Márquez-Luna, Gerardo Rodríguez-Tapia, and Víctor Moctezuma kindly provided photographs and maps. Adrián Fortino patiently corrected my figures and helped me improve them. Editor Lars Koerner and three anonymous reviewers provided very useful suggestions. For more than two decades, the Universidad Nacional Autónoma de México (UNAM) has generously provided me with a place to teach undergraduate and graduate students while doing research in systematics and biogeography with the most complete academic freedom. I am indebted to Mexico, my chosen homeland, which represents so many and sometimes contradictory things that cannot be expressed appropriately with words. At home, Adrián Fortino (*Homo sapiens*), Cocoa and Gamora (*Canis familiaris*), and Emma, Tiger, Leni, and Curly (*Felis catus*) have provided love and support.

Mexico City, Mexico Juan J. Morrone
March 16, 2020

Contents

1 What Is a Biogeographic Transition Zone? 1
1.1 Introduction 1
1.2 Biogeographic Transition Zones 5
1.3 Biotas, Cenocrons, and Horobiotas 7
1.4 Detection and Characterization of Transition Zones 8
1.5 Varieties of Biogeographic Transition Zones. 11
1.6 Biogeographic Hierarchy, Transition Zones, and Boundaries 13
1.7 Transition Zones of the World 15
References. 17

2 What Is Evolutionary Biogeography?. 21
2.1 Introduction 22
2.2 Identification of Biotas. 24
2.3 Testing Relationships Among Biotas. 35
2.4 Biogeographic Regionalization 44
2.5 Identification of Cenocrons 48
2.6 Construction of a Geobiotic Scenario 55
References. 57

3 A Historical Perspective of the Mexican Transition Zone. 69
3.1 Introduction 69
3.2 Halffter's Initial Contributions. 71
3.3 Conflicting Vertebrate and Insect Patterns. 76
3.4 The Mountain Entomofauna 79
3.5 Halffter's Distributional Patterns and Biotic Assembly. 81
3.6 The Mexican Transition Zone in the Twenty-First Century. 83
3.7 Impact of Halffter's Theory 85
References. 95

4 Biogeographic Regionalization of the Mexican Transition Zone 103
4.1 Introduction 104
4.2 Mexican Transition Zone 107
4.3 Sierra Madre Occidental Province 112
4.3.1 Apachian District 115
4.3.2 Durangoan District 116
4.4 Sierra Madre Oriental Province 118
4.4.1 Austral-Oriental Subprovince 122
4.4.2 Hidalgoan Subprovince 124
4.5 Transmexican Volcanic Belt Province 125
4.5.1 West Subprovince 130
4.5.2 East Subprovince 131
4.6 Sierra Madre del Sur Province 133
4.6.1 Western Sierra Madre del Sur Subprovince 136
4.6.2 Central Sierra Madre del Sur Subprovince 137
4.6.3 Eastern Sierra Madre del Sur Subprovince 138
4.7 Chiapas Highlands Province 140
4.7.1 Sierra Madrean District 143
4.7.2 Comitanian District 143
4.7.3 Lacandonian District 143
4.7.4 Soconusco District 144
4.7.5 Guatemalan Highland District 145
4.7.6 Nicaraguan Montane District 145
References 145

5 The Biotic Assembly of the Mexican Transition Zone 157
5.1 Biotic Assembly in a Biogeographic Transition Zone 157
5.2 The Paleoamerican Distributional Pattern: The Original Biota 158
5.3 The Mexican Plateau Cenocron: Old South American/Gondwanan Immigrants 162
5.4 The Mountain Mesoamerican Cenocron: Central (and South) American Immigrants 164
5.5 The Nearctic Cenocron: Northern Immigrants 168
5.6 The Typical Neotropical Cenocron: Last Southern Immigrants 170
5.7 Assembly of the Cenocrons in the Mexican Transition Zone: The Geobiotic Scenario 175
5.8 Relevance of the Biotic Assembly of the Mexican Transition Zone 179
References 180

6 Perspectives 185
6.1 Introduction 185
6.2 Evolutionary Biogeography 186
6.3 Ecological Biogeography 187
6.4 Integrative Biogeography 188
References 189

Chapter 1
What Is a Biogeographic Transition Zone?

Everything should be understood, and anything can be transformed—that is the modern view.

Susan Sontag (1992), The volcano lover

Abstract A biogeographic transition zone is a geographical area of overlap, with a gradient of replacement and partial segregation between different biotas (sets of taxa sharing a similar geographic distribution as a product of a common history). It is an area where physical features and environmental conditions allow the mixture and co-occurrence of species belonging to two or more biotas, but also constrain their distribution further into one another. The biogeographic affinities of the taxa assigned to these biotas are the most fundamental information considered to analyze accurately biogeographic transition zones. Ecological biogeographers have plotted the frequency of different distribution patterns on maps, detecting gradual changes in their relative contribution to a given area and identifying the most heterogeneous places in terms of distributional patterns as transition zones. Evolutionary biogeographers have found transition zones particularly interesting for analyzing causal connections between evolutionary and geological processes at large spatial and temporal scales. Biogeographic transition zones constitute natural laboratories for investigating evolutionary and ecological principles shaping biotic assembly. Additionally, they represent places where different evolutionary lineages coexist, having important implications for conservation, particularly when they also exhibit high diversity.

1.1 Introduction

The occurrence of different species and supraspecific taxa in particular geographical areas, known as Buffon's law, was noted since the eighteenth century (Morrone 2009). During the nineteenth and twentieth centuries, information on the

J. J. Morrone, *The Mexican Transition Zone*, https://doi.org/10.1007/978-3-030-47917-6_1

distributional patterns of plant and animal species accumulated, and eventually a worldwide picture emerged. The restriction of different plant and animal taxa to particular areas of the world allowed to recognize different phytogeographic, zoogeographic, and biogeographic units (e.g., Sclater 1858; Wallace 1876; Engler 1882; Takhtajan 1986; Moreira-Muñoz 2007; Holt et al. 2013; Morrone 2014a). These biogeographic units may be profitably analyzed in evolution and macroecology, to assess the degree of niche conservatism in different lineages over evolutionary time (Vilhena and Antonelli 2015).

In some phytogeographic and zoogeographic regionalizations of the world, clear differences between geographically distinct biotas were noticed, and kingdoms and regions were defined, although the precise delimitation of boundaries between them was quite elusive. One of the most striking examples of the difficulties in identifying such boundaries is the archipelago that separates the Oriental and Australian biogeographic regions, in southeastern Asia. This area was studied originally by Wallace (1860, 1863), who tried to establish the boundaries separating both zoogeographic regions as a line, but found a gradual transition, with animal species of different islands showing affinities to the Oriental or the Australian region and even to India and Africa. Zoogeographers soon became aware that biotas usually intergrade into one another as zones rather than lines, but chose to represent boundaries between biotic regions using lines on maps (Ferro and Morrone 2014). The complexity of such boundaries is evidenced when comparing alternative proposals by authors studying different taxonomic groups. For example, the original "Wallace line" (Wallace 1863) was modified by Murray (1866), Huxley (1868), Lydekker (1896), Sclater and Sclater (1899), and Mayr (1944), among others, all showing different breaks in an overall transition (Fig. 1.1). (For historical accounts of Wallace's line, see Camerini 1993 and van Oosterzee 1997.)

Biogeographic transition zones have not received the same attention than other biogeographic concepts. Although biogeographic regions and the transition zones between them are two different manifestations of the same phenomenon, the latter often remain as anecdotal within the framework of regionalization and without a parallel conceptual development (Ferro and Morrone 2014). Darlington (1957) included a section referring to transitions between regional faunas, where he stated that they are particularly complex and warned that his treatment of these zones was superficial. He defined a transition zone as the area where different faunal elements overlapped with subtractions in both directions. Pielou (1992) considered that transition zones have depauperate biotas because few elements from each region were found in the transition; however, other authors found that some transition zones may be extremely species-rich, such as the Mexican Transition Zone (Halffter 1987; Arita 1997; Ortega and Arita 1998). Morrone (2004) defined biogeographic transition zones as areas of biotic overlap (Fig. 1.2), promoted by historical and ecological changes that allow the mixture of taxa belonging to different biotas. Halffter and Morrone (2017) considered that transition zones are particularly important for evolutionary biogeography, because they allow to analyze the assembly of cenocrons with different taxonomic composition, dispersal capabilities, speciation modes, and ecological inertia.

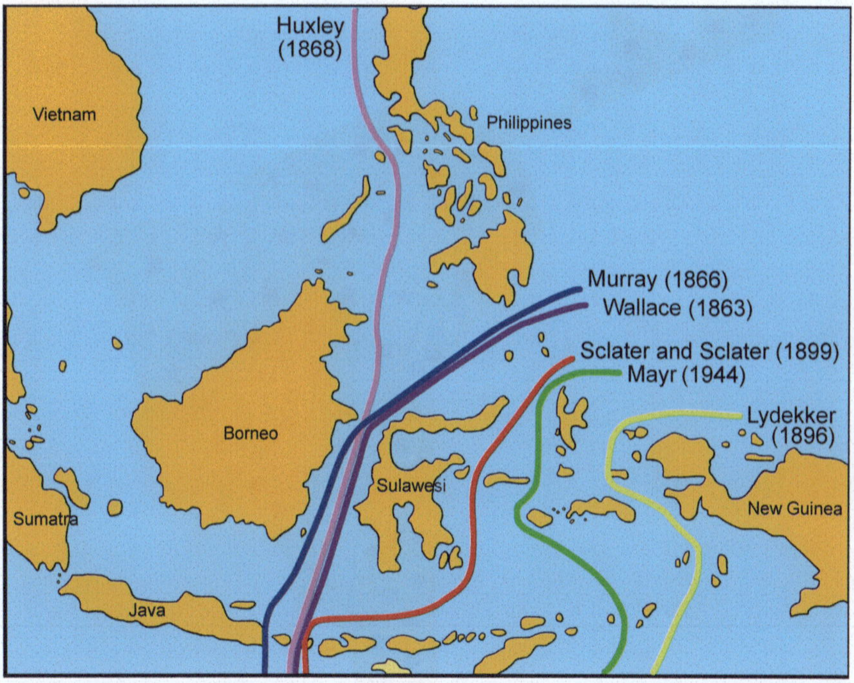

Fig. 1.1 Delimitation of Wallace's transition zone according to different authors

Biogeographic transition zones are specially relevant for analyzing biotic patterns and processes and to explore causal connections between biological and Earth history (Riddle and Hafner 2010). Wallace (1876) was one of the first biogeographers to realize their relevance, when acknowledging that, in addition to the overlapping distribution patterns, there were ongoing geological processes related to their development. During most of the twentieth century, authors dealing with transition zones of the world (e.g., Simpson 1940, 1965, 1977; Darlington 1957) emphasized dispersal as an explanation for the biotic assembly in the transition zone. Darlington (1957) postulated that wherever regional faunas overlap or are separated by partial barriers, a transition zone is established. Adjacent regional faunas consist of shared families, genera, and species; other taxa occur mostly in one region but extend in a part of the other; and some taxa occur in one region but not the other (Fig. 1.3). It was not until the last decades of the twentieth century (e.g., Reig 1981; Halffter 1987) that the relevance of vicariance was fully acknowledged as a contributing factor, leading to an evolutionary integrative approach (Morrone 2009). Recent advances in reconstructing Earth's history, molecular phylogenetics, phylogeography, and lineage dating, as well as understanding the integrative nature of biogeography, have provided evidence for a more accurate characterization of transition zones and for analyzing biotic assembly (Riddle and Hafner 2010).

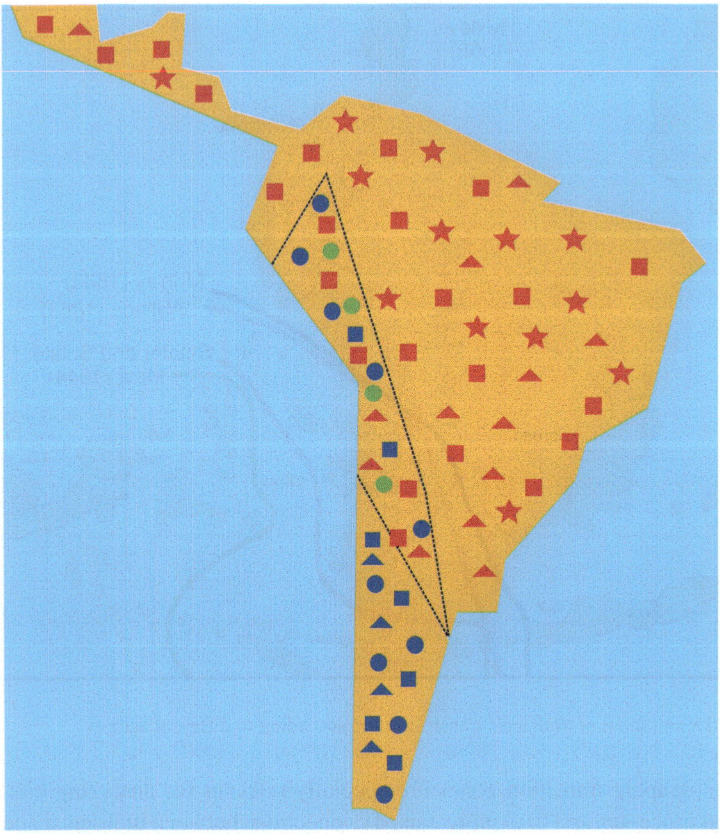

Fig. 1.2 Schematic representation of the South American Transition Zone. Red symbols represent Neotropical species; blue symbols represent Andean species; green symbols represent species endemic to the transition zone

Biogeographic transition zones generally refer to boundaries between biogeographic regions, but they may exist at other hierarchical levels such as subregions, provinces, or even districts (Morrone 2006). Furthermore, there may be different types of geographical transitions (physiographic, physiognomic, climatic, etc.). The differences and similarities between the different kinds of transition zones, as well as the interaction between them, might help address the artificial distinction between evolutionary and ecological biogeography. Ferro and Morrone (2014) considered that a conceptual synthesis might be possible, by trying to discover evolutionary and ecological principles ruling biogeographic transition zones at a variety of spatial and temporal scales.

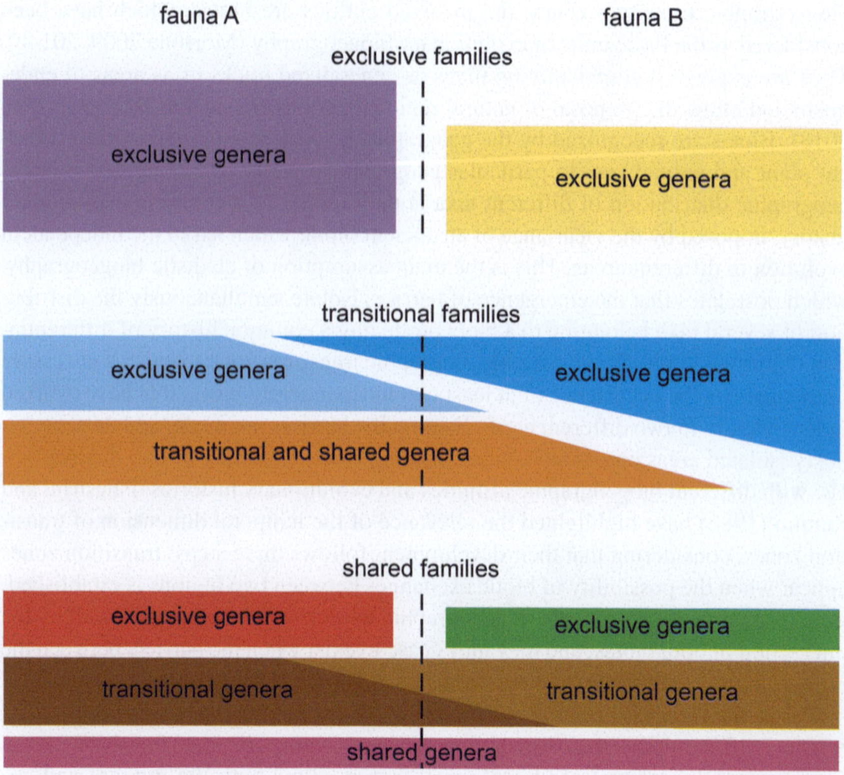

Fig. 1.3 Schematic representation of the transition between two biotas. Each biota consists of exclusive, transitional, and shared families; and transitional and shared families consist of exclusive, transitional, and shared genera (Modified from Darlington 1957)

1.2 Biogeographic Transition Zones

A transition is a passage from one form, state, or place to another (Merriam-Webster 2013). Thus, a transition requires the existence of at least two different entities that are connected. Ecological transitions may be identified over a broad spectrum of spatial and temporal scales (Gosz 1993). For example, the ecotone has been defined as a transition between two or more different communities (Odum 1971) or a zone of transition between two adjacent ecological systems (Holland 1988), among other definitions. The ecotone concept arose from community ecology to indicate a change in structure and composition of plant communities, but its use was then generalized to broader spatial scales as biomes (Risser 1995) or smaller scales as patches (Gosz 1993).

The specific features of a transition zone depend on the nature of the entities between which the transition occurs; for instance, classic definitions of ecological units involve mainly functional or structural criteria (Jax 2006). In the case of biogeographic transition zones, the involved entities are biotas, which have been considered as the basic units of evolutionary biogeography (Morrone 2009, 2014b). They are expressed graphically on maps as generalized tracks or as areas of endemism and allow the proposal of natural regionalizations (Escalante 2009; Morrone 2018). Biotas are recognized by the geographical restriction (endemism) of different plant and animal taxa to particular geographical areas. The congruence in the geographic distribution of different taxa is the product of a common evolutionary history, imposed by the vicariance of an ancient biota, which led to the independent evolution in different areas. This is the main assumption of cladistic biogeography, which postulates that the emergence of barriers isolate simultaneously the distribution of several taxa belonging to a biota producing a common history of differentiation (Morrone 2009). Thus, for a biogeographic transition zone to exist, a necessary prerequisite is the occurrence of at least two independently biotas that have evolved independently in two different areas. Eventually, barriers attenuate, and these previously isolated areas come into contact, leading to the assembly of two distinct biotas, with different biogeographic affinities and evolutionary histories. Palestrini and Zunino (1986) have highlighted the relevance of the temporal dimension of transition zones, considering that their development follows three steps: transition zones appear when the possibility of biotic exchanges between two regions is established; they evolve in response to the physiographic evolution of the area, as well as the interaction of both biotas; and they may cease to exist when the barriers between the regions are re-established.

Partial barriers (Darlington 1957) or filters (Simpson 1965; Rapoport 1975) restrict differentially the distribution of each biota in the transition zone. Environmental conditions and ecological factors allow both the mixture and co-occurrence of biotas that have different geographical origins, but also constrain their distribution further one into the other. The distributional restriction of such biotas may be a strong environmental gradient of unsuitable habitats (Glor and Warren 2010). For example, sharp environmental gradients may occur in transition zones associated with mountain ranges, as the Mexican Transition Zone, where temperature variation is crucial (Antonelli 2017; Rahbek et al. 2019). Paths of unsuitable habitats may have an underlying environmental gradient but not necessarily sharp; for example, in the case of the Indo-Malayan Transition Zone, in addition to the sea arms separating different islands, there is an aridity gradient between Sundaland and the Papuan area (Mayr 1944). The Sahara Desert, which represents the transition between the Palaearctic and Ethiopian (also known as Afrotropical) regions, has a gradient of aridity that seems to be a stronger barrier for passerine birds than the Mediterranean Sea (Rapoport 1975).

Whatever the kind of physical or environmental phenomena restricting species distribution of a given biota, the outcome is a more or less abrupt change in species composition of different taxonomic groups, which corresponds to a change in biogeographic affinities, in terms of present distribution and phylogenetic affinities, of the taxa involved. Partial barriers or filters do not affect exactly all species in the same way. For some species they may represent insurmountable barriers, other species may be not affected, and other species may be affected in different degrees.

Although physical and environmental phenomena restricting species distributions are prevalent all around the world, biogeographic transition zones, as considered so far only between biogeographic regions, occur in a few particular areas of the world. Therefore, there are historically contingent geological processes that are involved in the location of biogeographic transitions zones (Ferro and Morrone 2014).

1.3 Biotas, Cenocrons, and Horobiotas

From an evolutionary perspective, it is relevant to identify biogeographic units. There are several terms that have been applied to refer to these units, namely, elements, chorotypes, areas of endemism, and generalized tracks, among others (Reig 1981; Hausdorf 2002; Morrone 2014b; Passalacqua 2015; Fattorini 2016; Ferrari 2017). When analyzing the biotic assembly in transition zones, I find useful to distinguish between two different concepts: biota and cenocron (Morrone 2009, 2014b):

A biota corresponds to the living organisms of a region (Merriam-Webster 2013). The term "fauna" may be used to refer exclusively to animal taxa and the term "flora" to refer exclusively to plant taxa. There are several concepts that may be considered related to the term "biota," e.g., concrete biota, chronofauna, area of endemism, nuclear area, center of endemism, generalized track, biogeographic assemblage, taxonomic assemblage, and species assemblage (Morrone 2014b).

A cenocron refers to a set of taxa that share the same biogeographic history, which constitute an identifiable subset within a biota by their common biotic origin and evolutionary history (Morrone 2009). The term cenocron was proposed explicitly to refer to the dispersal and subsequent relatively synchronic implantation of a group of allochthonous taxa in a biota (Reig 1981). There are several concepts that incorporate a temporal dimension when implying the incorporation of taxa to a biota and may be considered similar to cenocron, e.g., biotic element, historical source, historical component, element, dispersal pattern, distributional pattern, lineage, and historical biota (Morrone 2014b).

Once a cenocron is incorporated to a biota, we may use the term horobiota to refer to the resulting assemblage. This term was defined by Reig (1981; as horofauna) as the set of species that coexist and diversify during an extended lapse and thus represent a lasting biogeographic unit. In this book I use this general term to refer to the different assemblages resulting from the dispersal of cenocrons to a transition zone (see Chap. 5).

The use of these terms can allow to account for patterns resulting from both vicariance and dispersal (Morrone 2014b). Biotas are the result of vicariance, which affects several taxa at the same time, whereas cenocrons are the result of dispersal, commonly geodispersal (Morrone 2009). These terms are relative: after the assembly of a cenocron into a biota, this "new" biota or horobiota may behave in the future as a cenocron in relation to another biota. Instead of assuming dispersal or vicariance as the only driver of biotic assembly, the dispersal-vicariance model

(Brooks 2004; Lieberman 2004; Morrone 2009) considers both processes to be relevant.

1.4 Detection and Characterization of Transition Zones

Not all the species inhabiting a transition zone are affected exactly in the same way by partial barriers or filters. Thus, a transition zone is an area of overlap with differential penetration of taxa from one biota into another. Depending on the nature of the barrier and the taxon under study, transition zones may vary from narrow zones with strong changes in biotic composition to broad zones with gradual biotic changes along their length (Ferro and Morrone 2014). Irrespective of the nature of the barrier and considering either one taxon or the whole biota, a transition zone involves an area with a gradient of biotic change. The lines drawn on maps by early naturalists at the boundaries between major biogeographic regions are useful as easily transmissible syntheses that indicate changes in biotic composition associated with biogeographic transitions zones; however, these lines fall within a zone of replacement gradients, where each author considers is located the strongest biotic interchange. Associational networks (Vilhena and Antonelli 2015) abstract species presence-absence distributional data as networks, incorporating complex relationships instead of similarity measures, where regions appear as highly interconnected groups of localities. Vilhena and Antonelli (2015) compared the performance of the species turnover and network approaches with a simulated data set (Fig. 1.4a), finding that the biogeographic transition zone may be engulfed by one of the regions when two clusters are chosen and it may represent a distinct region if three clusters are chosen (Fig. 1.4b). When they applied the network method to the same data, four clusters were found (Fig. 1.4c): one with cells 1–14, another with cells 17–30, and grid-cells 15 and 16 each forming their own clusters.

In evolutionary biogeography, transition zones may be detected by the presence of panbiogeographic nodes, namely, areas where different generalized tracks converge (Morrone 2009, 2018). These nodes point out places where biotic assembly occurs; however, they do not help distinguish the width of a transition zone (Miguel-Talonia and Escalante 2013). They are usually found in biogeographic provinces that are denoted as transitional or in the boundaries between different provinces (Escalante et al. 2004; Morrone and Márquez 2008). In cladistic biogeographic analyses, transition zones may be detected by conflicting results, where a putative transition zone may result to be the sister area to different biogeographic areas (Morrone 2009). Cladistic biogeographic analyses are based on predefined areas of endemism; thus, transition zones are represented on a general area cladogram by specific areas of endemism that have hybridized. This approach detects areas of endemism as transitional, with a defined extension and boundaries, so that the separation between the regions may be seen as a clearly defined area, in contrast with the nodes detected by track analyses. Thus, track analysis and cladistic biogeography capture different features of the transition zone (Ferro and Morrone 2014).

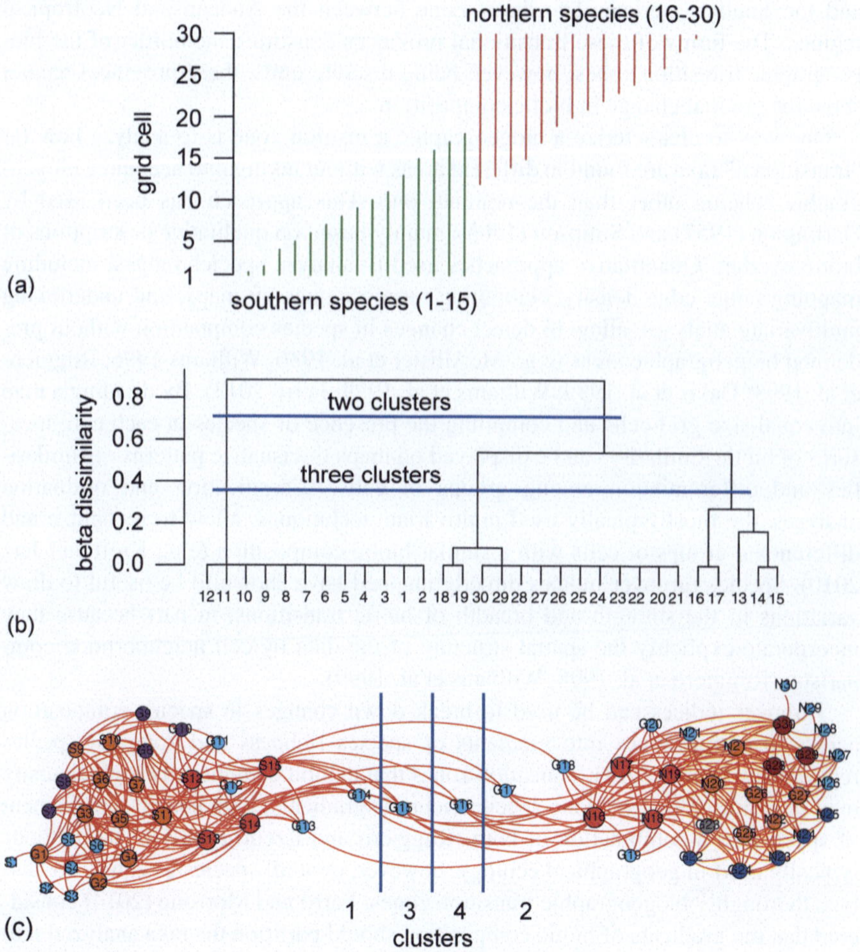

Fig. 1.4 Detection of a transition zone using species turnover and network approaches. (**a**) Species range data across 30 grid-cells; data represent 2 biogeographic regions that overlap in a transition zone; (**b**) clustering these data with an unweighted pair group method, 2 or 3 clusters are obtained, where 3 clusters cause the transition zone to appear as a distinct region; (**c**) in the network clustering, the optimal representation is 4 clusters, where the transition zone is composed of 2 clusters, each containing a single species that cannot be confidently assigned to any of the major regions (Modified from Vilhena and Antonelli 2015)

The width of a transitional zone is variable, depending on the author's criteria. For example, Wallace (1876) considered whole subregions as transitional between biogeographic regions. Despite this, the border line of a transitional biogeographic unit assigned to a biogeographic region is usually drawn as the limit of that biogeographic region. Morrone (2006) analyzed the biogeographic regionalization of the New World and defined two groups of provinces as transitional zones between its regions: the Mexican Transition Zone between the Nearctic and Neotropical regions and the South American Transition Zone between the Andean and Neotropical regions. The limits of these transitional provinces constitute the border of the biogeographic transition zones; however, being discrete units, these provinces cannot show the gradual change in biotic composition.

One way to characterize a biogeographic transition zone is to analyze how far "transitional" taxa are found in different areas without taking into account a biogeographic scheme other than the regional one. This approach has been used by Darlington (1957) and Simpson (1965), mainly based on qualitative descriptions of biotic overlap. Quantitative approaches used to analyze species ranges, including mapping range edge density, computing turnover rates on maps, and undertaking multivariate analyses, allow to detect changes in species composition without predefined biogeographic areas (e.g., McAllister et al. 1986; Williams 1996; Ruggiero et al. 1998; Davis et al. 1999; Williams et al. 1999; Ferro 2013). By dividing a map into equal size grid-cells and compiling the presence of species in each cell, measures of biotic similarity can be displayed on maps to visualize patterns of similarities and differentiation among groups of cells. Classification and ordination analyses, the most typically used multivariate techniques, allow to recognize and differentiate groups of cells with a similar biotic composition (e.g., Kreft and Jetz 2010). Species turnover indices directly mapped have shown to be useful to draw variations in the strength and breadth of biotic transitions, in part because they incorporate explicitly the spatial structure of the data by cell neighborhood comparison (Ruggiero et al. 1998; Williams et al. 1999).

Turnover indices can be used to break down changes in species composition across transition zones into gradients of species richness and zones of species replacement (Ferro 2013). Transition zones that exhibit an unusually high diversity may be represented by strong species richness gradients, high spatial replacement of species, or a combination of both (Ruggiero and Ezcurra 2003). The methods typically used in geographical ecology, however, treat all species as equal. To analyze thoroughly biogeographic transition zones, Ferro and Morrone (2014) considered that the gradients of biotic composition should partition the taxa analyzed into cenocrons. Thus, taxa assigned to different cenocrons should have different gradients of biotic composition.

Distributional patterns are fundamental for the analysis of biogeographic transition zones. Since shared distributional patterns are the basis of biogeographic regionalizations, the biogeographic affinities of taxa are the most fundamental information to consider in order to decompose accurately biogeographic transition zones (Ferro et al. 2017). The simplest way to define the biogeographic affinity of a given taxon is to recognize its range concordance to predefined geographical areas, such as continents in a regional-level regionalization. A more accurate way is to disaggregate range concordance according to smaller geographic areas nested within larger ones. This may generate a greater number of distributional patterns, but may allow a finer definition of their integration in a biogeographic transition zone. A quantitative approach to the definition of distributional patterns may be the identification of chorotypes, namely, the statistically significant groups of taxa with coincident distribution areas (Zunino 2005; Olivero et al. 2011; Ferro et al. 2017).

Another aspect of biogeographic affinity is the evolutionary history of the taxa, which can be inferred through phylogenetic analyses, for instance, by analyzing the distribution of their sister taxa (Brundin 1966; Palestrini and Zunino 1986; Roig-Juñent 1992) or by the optimization of the geographical distribution onto the cladogram (Ferro et al. 2017). Phylogenetic analyses of entire biotas would be required to assess their biogeographic affinities and evolutionary history, but raw distributional data can be used as an adequate preliminary approximation (Morrone 2001, 2009).

From an evolutionary biogeographic perspective, the recognition of cenocrons has been the most classic approach used to analyze transition zones (Halffter 1987; Halffter and Morrone 2017). Considered in a hypothetico-deductive framework, cenocrons represent testable hypotheses (Lobo 2007; Morrone 2015a; Halffter and Morrone 2017). There are different ways to falsify them, for example, dating selected lineages and examining their phylogenetic placement and the distribution of their related taxa. On the other hand, new analyses may help identify other cenocrons or refine those that have been already described. The precise identification of cenocrons allows to analyze more accurately the influence of vicariance events, helping discern cases when different area cladograms show the same area relationships although taxa diversified at different times (pseudo-congruence) and cases when area cladograms show conflict but the age of the taxa indicates that they diversified in response to different events (pseudo-incongruence).

If hypotheses on cenocrons are available for a given area, it would be possible to undertake a time-sliced cladistic biogeographic analysis (Cecca et al. 2011). For example, in a hypothetical case where two cenocrons have dispersed to the biota distributed in a given area, three different time-slices may be identified. The oldest time-slice corresponds only to the taxa belonging to the original biota, the intermediate time-slice corresponds to the taxa belonging to the original biota + the taxa of the first cenocron, and the most recent time-slice corresponds to all the taxa together. Amorim et al. (2009) refer to taxa that belong to different cenocrons and inhabit the same time-slice as "allochronic taxa." The separate cladistic biogeographic analyses for these three time-slices could help understand the way vicariance has affected these successive horobiotas (Corral-Rosas and Morrone 2017).

1.5 Varieties of Biogeographic Transition Zones

The simplest approach to analyze different distributional patterns in a given biogeographic transition zone is to divide its taxa by their biogeographic affinities in four different sets (Ferro and Morrone 2014): (1) taxa distributed in one region and the transition zone; (2) taxa distributed in the other region and the transition zone; (3) widespread taxa distributed in both regions and the transition zone; and (4) taxa endemic to the transition zone. The spatial arrangement of these basic distributional patterns in a transition zone may generate either depauperate or species-rich

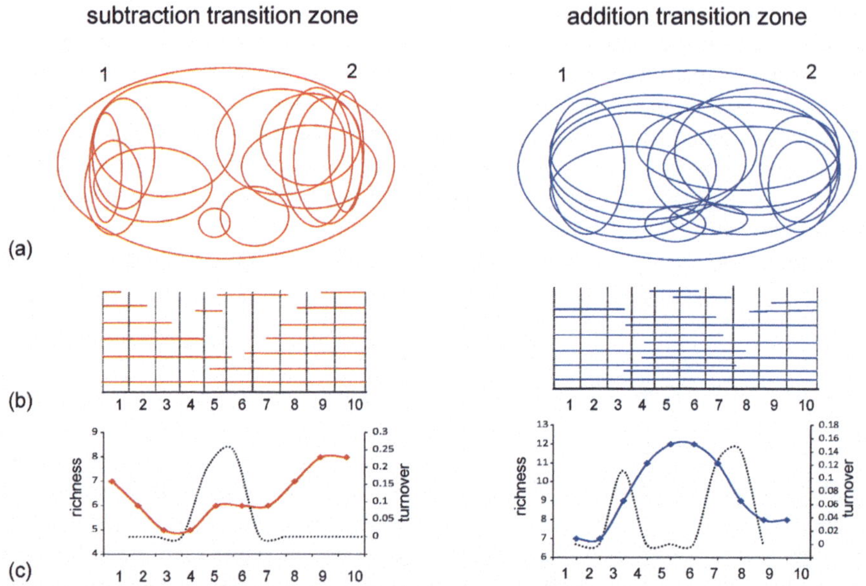

Fig. 1.5 Subtraction and addition transition zones. (**a**) Hypothetical distributional areas; two sets of geographically contiguous biotas, plus one widespread and one range-restricted endemic biota; (**b**) one-dimensional representation of range overlap; the level of juxtaposition generates subtraction or addition transition zones; (**c**) solid lines are the richness gradients; dotted lines are turnover values, where the highest turnover rate can occur at either the center (subtraction transition zone) or the sides (addition transition zone) (Modified from Ferro and Morrone 2014)

transition zones. Based on the level of juxtaposition of transitional elements, two types of transition zones may be distinguished (Fig. 1.5):

Subtraction Transition Zones: they show low overlap of biotas and progressive loss of taxa when passing from one region to the other. These transition zones are expected to be depauperate as a consequence of the progressive loss of species from both regions. One example is the Saharo-Arabian Transition Zone.

Addition Transition Zones: they show high overlap of biotas and progressive gain of taxa of each region when passing from one region to the other. They are expected to be species-rich transition zones. One example is the Mexican Transition Zone.

Variations of these oversimplified hypothetical cases may occur at different hierarchical levels of a biogeographic regionalization.

1.6 Biogeographic Hierarchy, Transition Zones, and Boundaries

Biogeographic units are biological systems organized hierarchically, whereas smaller units are nested within larger ones producing emergent properties. Biogeographic regionalization has a hierarchical organization where districts are nested within provinces, these are nested within dominions, these in regions and the latter in kingdoms (Ebach et al. 2008; Escalante 2009; Morrone 2018). The hierarchy of the taxonomic categories determines that the geographical distributions of taxa are also ordered hierarchically; the geographical distribution of a family contains the distribution of all its genera and species; however, the biogeographic hierarchy is not directly related to the taxonomic hierarchy. For instance, the distribution of a species may correspond to a whole biogeographic region, or a family may be endemic to a province. Because evolution is a dynamic process, the geographical surrogates we perceive and define as areas of endemism are the result of the continuous shaping of geological, climatic, and ecological contingencies through historical time. For example, substantial diversification of some lineages may occur by adaptive radiation defining geographical and taxonomically consistent biogeographic units, while other relictual ancient lineages may occur in rather small geographical areas, obscuring the relationship between the taxonomical and biogeographic hierarchies. The broad correspondence between the hierarchy of natural groups and the hierarchy of biogeographic regions is one of the most persuasive evidence of the influence of geography and isolation of biotas in shaping evolution (Ferro and Morrone 2014). Thus, it is likely that the longer the evolutionary history of an isolated biota in a given area, the more time for taxa of higher categories (e.g., orders or families) to diversify and thus, the more likely they define higher hierarchy biogeographic units (e.g., realms or regions). Consequently, the finest biogeographic subdivisions, the districts, reflect a more recent history of isolation with probably less marked barriers and more recently diversified lineages (e.g., species).

The taxonomic contrasts in successively finer biogeographic divisions are obviously less sharp than in regional-level transitions. Therefore, it is not surprising that biogeographers have focused on these broad-level transitions zones. On the other hand, ecologists, by means of experimental and comparative procedures at local spatial and temporal scales, aimed to explain transition zones mostly in terms of actual environmental conditions and ecological interactions (e.g., Fergnani et al. 2013). Different methodological and conceptual frameworks for the analysis of species diversity, such as focusing in explaining the origin or the maintenance of diversity, have led to a progressive divergence between ecological and evolutionary biogeography since the early twentieth century (Ricklefs and Jenkins 2011). Both perspectives, however, are not as independent as have been assumed in the last century. The differences between ecological and evolutionary biogeography are just a matter of scales, both temporal and spatial, of a single continuum known as biological evolution observed in a geographical context (Morrone 2009). The fact that most biogeographic transition zones coincide with areas of change in environmental gradients illustrates this interaction (Ferro and Morrone 2014). Their position and amplitude are the result of complex spatial-temporal interaction between contemporary and past climatic and geomorphological features. Environmental gradients play an important role in maintaining the isolation between biotas, acting as ecological barriers that limit the spread of the species, and even creating, over evolutionary time periods, consistent geographical distribution patterns without the presence of an evident physiographic barrier (Endler 1977). These processes are clearly not mutually exclusive and act jointly.

As one moves down in the biogeographic hierarchy, the difference between ecological and evolutionary biogeography turns fuzzy, and both subdisciplines blend (Ferro and Morrone 2014). The smallest areas of endemism recognized in a regionalization, known as districts, frequently coincide with bioclimatic zonation and life form zonation. Thus, small areas of endemism have their own entity, as geographic evolutionary units, and, therefore, may as well exhibit transition zones, if contiguous, when passing from one entity to another. From a strictly ecological perspective, terms as ecotone, ecocline, interface, edge, gradient, border, and transition zone have been applied to describe the passage between communities, biomes, or ecological systems, encompassed under the more inclusive concept of ecological boundaries (Yarrow and Marín 2007). An ecological boundary has been defined as "a zone of transition between contrasting systems with a gradient in the feature setting up the contrast steeper in the boundary than in adjoining systems and a wideness or narrowness of the boundary reflecting the steepness of the gradient" (Cadenasso et al. 2003, p. 718).

The concept of biogeographic transition zone may be related to the concept of ecological boundary. Ecological boundaries act as physical filters, like a semipermeable membrane, controlling the quantity and quality of energy and material flux across their interface (Strayer et al. 2003), and may occur at any level of the ecological hierarchy. The ecological hierarchy includes entities involved in the transfer of matter and energy, known as interactors, namely, molecules, cells, organisms, populations, communities, and biotas. The genealogical hierarchy includes entities known as replicators, namely, genes, chromosomes, organisms, populations, species, and clades, that may reproduce into similar entities and evolve (Morrone 2004). Organisms and populations are common to both hierarchies and may be seen as either interactors or replicators. Ecological boundaries may have repercussions at the level shared with the genealogical hierarchy, in controlling the flux of information. Thus, they may have an important role in shaping the geographic distribution of replicators.

It is important to note that ecological boundaries or ecotones do not always represent biogeographic transition zones as defined herein. For instance, differences in dominance of some species or sets of characteristic species across environmental gradients are frequently described in the ecological literature (Gosz 1993). These ecotones can be seen as ongoing processes of differentiation (Schneider et al. 1999), but not as biogeographic transition zones unless endemism of several taxa occur. Therefore, biogeographic transition zones may occur in lower hierarchical level of the biogeographic regionalization, such as districts, as long as at least two biotas, in turn defined by at least two sets of endemic species, get into contact geographically.

1.7 Transition Zones of the World

Five main biogeographic transition zones (Fig. 1.6) have been recognized for the world (Morrone 2015b):

Mexican Transition Zone: it includes the mountainous areas of Mexico, Guatemala, Honduras, El Salvador, and Nicaragua north of Lake Nicaragua (Morrone 2014a, 2015b; Halffter and Morrone 2017). It corresponds to the boundary between the Nearctic and Neotropical regions and is comprised of the Sierra Madre Occidental, Sierra Madre Oriental, Sierra Madre del Sur, Transmexican Volcanic Belt, and Chiapas Highlands (Morrone 2014a, 2015a, b). Halffter (1987, 2017) considers that the Mexican Transition Zone extends to the Southern United States as well as the Mexican lowlands (see Chap. 3).

Saharo-Arabian Transition Zone: it comprises the Sahara Desert and the Arabian Peninsula (Müller 1986; Kreft and Jetz 2013). Some authors extend its eastern

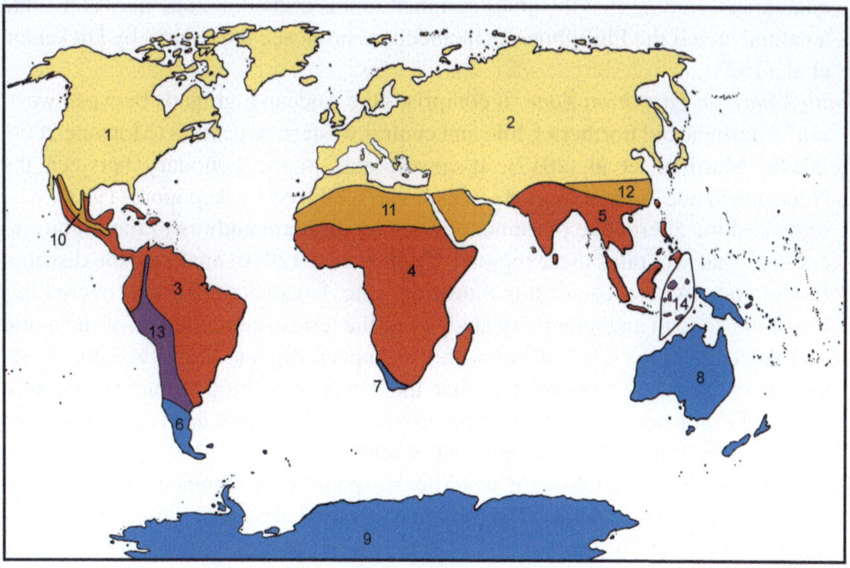

Fig. 1.6 World biogeographic regionalization, with indication of the regions and transition zones. (1) Nearctic region; (2) Palearctic region; (3) Neotropical region; (4) Ethiopian region; (5) Oriental region; (6) Andean region; (7) Cape region; (8) Australian region; (9) Antarctic region; (10) Mexican Transition Zone; (11) Saharo-Arabian Transition Zone; (12) Chinese Transition Zone; (13) South American Transition Zone; (15) Indo-Malayan Transition Zone (Modified from Morrone 2015b)

limits to western Pakistan, naming it Saharo-Sindian Transition Zone (Mario Zunino pers. comm.). It corresponds to the boundary between the Palearctic and Ethiopian regions. Müller (1986) provided some examples of taxa from this transition zone.

Chinese Transition Zone: it corresponds to the boundary between the Palearctic and Oriental regions (Palestrini et al. 1985; Müller 1986; Kreft and Jetz 2013). Müller (1979) suggested that this zone extends from the Yang Tsê-Kiang River to the 21° parallel, including also Taiwan, based on the distribution of different taxa. Palestrini et al. (1985) analyzed the geographical distribution of some groups of Scarabaeoidea (Coleoptera) of this area and detected the overlap of Palearctic, Oriental, and Sino-Japanese cenocrons.

Indo-Malayan or Indonesian Transition Zone: it is also known as Wallacea and corresponds to the boundary between the Oriental and Australian regions (Dickerson et al. 1923; Mayr 1944; Darlington 1957; Simpson 1977; Müller 1986; Vallejo 2011; Kreft and Jetz 2013). Müller (1986) discussed its boundaries and gave examples of Oriental and Australian taxa overlapping in this transition zone. Michaux (2010) analyzed the geological development of this area, identified areas of endemism, and concluded that the latter are linked to geological processes resulting from the interaction between the Eurasian and Australian continents and the Philippine Sea Plate. King and Ebach (2017) showed that it is a temporally composite area. Michaux (2019) undertook a parsimony analysis of endemicity, finding that the areas assigned to this transition zone may constitute a natural area if the Philippines is included, as proposed previously by Dickerson et al. (1923).

South American Transition Zone: it comprises the Andean highlands between western Venezuela and northern Chile and central western Argentina (Morrone 2006, 2014a; Martínez et al. 2017). It corresponds to the boundary between the Neotropical and Andean regions, which was analyzed by Rapoport (1968), who discussed the alternative placements given by different authors to the "subtropical line" that separates these regions. Urtubey et al. (2010) analyzed the distribution of some Asteraceae of this transition zone. Escalante (2017) recovered this transition zone in an endemicity analysis of the terrestrial mammals of the world and noted its close relationship with the Andean region. More recently, Roig-Juñent et al. (2018) considered that the Patagonian biogeographic province should be considered as belonging to the South American Transition Zone instead of the Andean region in the strict sense.

South African Transition Zone: it would correspond to the boundary between the Ethiopian and Cape regions. The presence of varied groups in South Africa (e.g., Verboom et al. 2009; Daru et al. 2016) may point to the existence of such area, although its proper demarcation has not been attempted yet.

References

Amorim DS, Santos CMD, Oliveira SS (2009) Allochronic taxa as an alternative model to explain circumantarctic disjunctions. Syst Entomol 34:2–9

Antonelli A (2017) Drivers of bioregionalization. Nat Ecol Evol 1:1–2

Arita HT (1997) The non-volant mammal fauna of Mexico: species richness in a megadiverse country. Biodivers Conserv 6:787–795

Brooks DR (2004) Reticulations in historical biogeography: the triumph of time over space in evolution. In: Lomolino MV, Heaney LR (eds) Frontiers of biogeography: new directions in the geography of nature. Sinauer Associates, Sunderland, MA, pp 111–124

Brundin L (1966) Transantarctic relationships and their significance as evidenced by midges. Kungl Sven Vetensk Akad Handl (ser 4) 11:1–472

Cadenasso ML, Picket STA, Weathers KC, Bell SS, Benning TL, Carreiro M, Dawson TE (2003) An interdisciplinary and synthetic approach to ecological boundaries. Bioscience 53:717–722

Camerini JR (1993) Evolution, biogeography, and maps: an early history of Wallace's line. Isis 84:700–727

Cecca F, Morrone JJ, Ebach MC (2011) Biogeographic convergence and time-slicing: concepts and methods in cladistic biogeography. In: Upchurch P, McGowan A, Slater C (eds) Palaeogeography and palaeobiogeography: biodiversity in space and time. Systematics Association Special Volume. Taylor and Francis, CRC Press, Boca Raton, FL, pp 1–12

Corral-Rosas V, Morrone JJ (2017) Analyzing the assembly of cenocrons in the Mexican transition zone through a time-sliced cladistic biogeographic analysis. Aust Syst Bot 29:489–501

Darlington PJ Jr (1957) Zoogeography: the geographical distribution of animals. Wiley, New York

Daru BH, van der Bank M, Maurin O, Yessoufou K, Schaefer H, Slingsby JA, Davies TJ (2016) A novel phylogenetic regionalization of phytogeographical zones of southern Africa reveals their hidden evolutionary affinities. J Biogeogr 43:155–166

Davis ALV, Scholtz CH, Chown SL (1999) Species turnover, community boundaries and biogeographic composition of dung beetle assemblages across an altitudinal gradient in South Africa. J Biogeogr 26:1039–1055

Dickerson RE, Merrill ED, McGregor RC, Schultze W, Taylor EH, Herre AW (1923) Distribution of life in the Philippines. Manila Bur Sci Mon 21:1–322

Ebach MC, Morrone JJ, Parenti LR, Viloria AL (2008) International code of area nomenclature. J Biogeogr 35:1153–1157

Endler JA (1977) Geographic variation, speciation and clines, Monographs in population biology, vol 10. Princeton University Press, Princeton

Engler A (1882) Versuch einer Entwicklungsgeschichte der Pflanzenwelt, insbesondere der Florengebiete seit der Tertiärperiode. vol 2. Die extratropischen Gebiete der Südlichen Hemisphäre und die Tropischen Gebiete. Verlag von W. Engelmann, Leipzig

Escalante T (2009) Un ensayo sobre regionalizaci6n biogeográfica. Rev Mex Biodivers 80:551–560

Escalante T (2017) A natural regionalization of the world based on primary biogeographic homology of terrestrial mammals. Biol J Linn Soc 120:349–362

Escalante T, Rodríguez G, Morrone JJ (2004) The diversification of Nearctic mammals in the Mexican Transition Zone. Biol J Linn Soc 83:327–339

Fattorini S (2016) A history of chorological categories. Hist Philos Life Sci 38:1–21

Fergnani PN, Sackmann P, Ruggiero A (2013) The spatial variation in ant species composition and functional groups across the Subantarctic-Patagonian transition zone. J Ins Conserv 17:295–305

Ferrari A (2017) Biogeographical units matter. Aust Syst Bot 30:391–402

Ferro I (2013) Rodent endemism, turnover and biogeographic transitions on elevation gradients in the northwestern Argentinian Andes. Mamm Biol 78:322–331

Ferro I, Morrone JJ (2014) Biogeographic transition zones: a search for conceptual synthesis. Biol J Linn Soc 113:1–12

Ferro I, Navarro-Sigüenza AG, Morrone JJ (2017) Biogeographic transitions in the Sierra Madre Oriental, Mexico, shown by chorological and evolutionary biogeographic affinities of Passerine birds (Aves: Passeriformes). J Biogeogr 44:2145–2160
Glor RE, Warren D (2010) Testing ecological explanations for biogeographic boundaries. Evolution 65:673–683
Gosz JR (1993) Ecotone hierarchies. Ecol Appl 3:369–376
Halffter G (1987) Biogeography of the montane entomofauna of Mexico and Central America. Annu Rev Entomol 32:95–114
Halffter G (2017) La Zona de Transición Mexicana y la megadiversidad de México: Del marco histórico a la riqueza actual. Dugesiana 24:77–89
Halffter G, Morrone JJ (2017) An analytical review of Halffter's Mexican transition zone, and its relevance for evolutionary biogeography, ecology and biogeographic regionalization. Zootaxa 4226:1–46
Hausdorf B (2002) Units in biogeography. Syst Biol 51:648–652
Holland MM (1988) SCOPE/MAB technical consultations on landscape boundaries. In: Di Castri F, Hansen AJ, Holland MM (eds) A new look at ecotones: emerging international projects on landscape boundaries, Biological International, Paris (special issue), vol 17, pp 47–106
Holt BG, Lessard JP, Borregaard MK, Fritz SA, Araújo MB, Dimitrov D, Fabre PH, Graham CH, Graves GR, Jønsson KA, Nogués-Bravo D, Wang Z, Whittaker RJ, Fjeldså RJ, Rahbek C (2013) An update of Wallace's zoogeographic regions of the world. Science 339:74–78
Huxley TH (1868) On the classification and distribution of the Alectoromorphae and Heteromorphae. Proc Zool Soc Lond 1868:296–319
Jax K (2006) Ecological units: definitions and application. Q Rev Biol 81:237–258
King AR, Ebach M (2017) A novel approach to time-slicing areas within biogeographic-area classifications: Wallacea as an example. Aust Syst Bot 30:495–512
Kreft H, Jetz W (2010) A framework for delineating biogeographic regions based on species distributions. J Biogeogr 37:2029–2053
Kreft H, Jetz W (2013) Comment on "An update of Wallace's zoogeographic regions of the world". Science 341:343
Lieberman BS (2004) Range expansion, extinction, and biogeographic congruence: a deep time perspective. In: Lomolino MV, Heaney LR (eds) Frontiers of biogeography: new directions in the geography of nature. Sinauer Associates, Sunderland, MA, pp 111–124
Lobo JM (2007) Los "patrones de dispersión" de la fauna ibérica de Scarabaeinae (Coleoptera). In: Zunino M, Melic A (eds) Escarabajos, diversidad y conservación biológica. Ensayos en homenaje a Gonzalo Halffter. Sociedad Entomológica Aragonesa, Monografías 3er. Milenio M3M, Zaragoza, pp 159–177
Lydekker R (1896) A geographical history of mammals. Cambridge University Press, Cambridge
Martínez GA, Arana MD, Oggero AJ, Natale ES (2017) Biogeographic relationships and new regionalisation of high-altitude grasslands and woodlands of the central Pampean Ranges (Argentina), based on vascular plants and vertebrates. Aust Syst Bot 29:473–488
Mayr E (1944) Wallace's Line in the light of recent zoogeographic studies. Q Rev Biol 19:1–14
McAllister DE, Platania SP, Schueler FW, Baldwin ME, Lee DS (1986) Ichthyofaunal patterns on a geographic grid. In: Hocutt CH, Wiley EO (eds) Zoogeography of North American freshwater fishes. Wiley, New York, pp 17–51
Merriam-Webster (2013) Merriam-Webster online dictionary. http://www.merriam-webster.com
Michaux B (2010) Biogeography of Wallacea: geotectonic models, areas of endemism, and natural biogeographic units. Biol J Linn Soc 101:193–212
Michaux B (2019) Biogeology: evolution in a changing landscape. CRC Press, Boca Raton, FL
Miguel-Talonia C, Escalante T (2013) Los nodos: El aporte de la panbiogeografía al entendimiento de la biodiversidad. Biogeografía 6:30–42
Moreira-Muñoz A (2007) The Austral floristic realm revisited. J Biogeogr 34:1649–1660
Morrone JJ (2001) Homology, biogeography and areas of endemism. Divers Distrib 7:297–300

Morrone JJ (2004) La Zona de Transición Sudamericana: Caracterización y relevancia evolutiva. Acta Entomol Chilena 28:41–50

Morrone JJ (2006) Biogeographic areas and transition zones of Latin America and the Caribbean Islands, based on panbiogeographic and cladistic analyses of the entomofauna. Annu Rev Entomol 51:467–494

Morrone JJ (2009) Evolutionary biogeography: an integrative approach with case studies. Columbia University Press, New York

Morrone JJ (2014a) Biogeographic regionalization of the Neotropical region. Zootaxa 3782:1–110

Morrone JJ (2014b) On biotas and their names. Syst Biodivers 12:386–392

Morrone JJ (2015a) Halffter's Mexican transition zone (1962-2014), cenocrons and evolutionary biogeography. J Zool Syst Evol Res 53:249–257

Morrone JJ (2015b) Biogeographic regionalization of the world: a reappraisal. Aust Syst Bot 28:81–90

Morrone JJ (2018) The spectre of biogeographic regionalization. J Biogeogr 45:282–288

Morrone JJ, Márquez J (2008) Biodiversity of Mexican terrestrial Arthropods (Arachnida and Hexapoda): a biogeographic puzzle. Acta Zool Mex (n s) 24:15–41

Müller P (1979) Introducción a la biogeografía. Blume, Madrid

Müller P (1986) Biogeography. Harper and Row, New York

Murray A (1866) The geographic distribution of mammals. Day and Son, London

Odum EP (1971) Fundamentals of ecology. WB Saunders, Philadelphia, PA

Olivero J, Real R, Márquez AL (2011) Fuzzy chorotypes as a conceptual tool to improve insight into biogeographic patterns. Syst Biol 60:645–660

Ortega J, Arita HT (1998) Neotropical-Nearctic limits in Middle America as determined by distributions of bats. J Mammal 79:772–783

Palestrini C, Zunino M (1986) L'analisi dell'entomofauna nelle zone de transizione: prospective e problemi. Biogeographia 12:11–25

Palestrini C, Simonis A, Zunino M (1985) Modelli di distribuzione dell'entomofauna della Zona di Transizione Cinese, analisi di esempi e ipotesi sulle sue origini. Biogeographia 11:195–209

Passalacqua NG (2015) On the definition of element, chorotype and component in biogeography. J Biogeogr 42:611–618

Pielou EC (1992) Biogeography. Krieger Publishing, Malabar

Rahbek C, Borregaard MK, Colwell RK, Dalsgaard B, Holt BG, Morueta-Holme N, Nogues-Bravo D, Whittaker RJ, Fjeldsa J (2019) Humboldt's enigma: What causes global patterns of mountain biodiversity? Science 365:1108–1113

Rapoport EH (1968) Algunos problemas biogeográficos del Nuevo Mundo con especial referencia a la región Neotropical. In: Delamare Debouteville C, Rapoport EH (eds.), Biologie de l'Amerique Australe, vol 4. Editions du Centre National de la Recherche Scientifique, Paris, pp 55–110

Rapoport EH (1975) Areografía: Estrategias geográficas de las especies. Fondo de Cultura Económica, Mexico City

Reig OA (1981) Teoría del origen y desarrollo de la fauna de mamíferos de América del Sur. Museo Municipal de Ciencias Naturales Lorenzo Scaglia, Mar del Plata

Ricklefs RE, Jenkins DG (2011) Biogeography and ecology: toward the integration of the two disciplines. Philos Trans R Soc B 366:2438–2448

Riddle BR, Hafner DF (2010) Integrating pattern with process at biogeographic boundaries: the legacy of Wallace. Ecography 33:321–325

Risser PG (1995) The status of the science examining ecotones. Bioscience 45:318–325

Roig-Juñent S (1992) Insectos de América del sur, su origen a través del enfoque de la biogeografía histórica. Multequina (Mendoza) 1:107–114

Roig-Juñent SA, Griotti M, Domínguez MC, Agrain FA, Campos-Soldini P, Carrara R, Cheli G, Fernández-Campón F, Flores GE, Katinas L, Muzón JR, Neita-Moreno JC, Pessacq P, San Blas G, Scheibler EE, Crisci JV (2018) The Patagonian Steppe biogeographic province: Andean region or South American transition zone? Zool Scripta 47:623–629

Ruggiero A, Ezcurra C (2003) Regiones y transiciones biogeográficas: Complementariedad de los análisis en biogeografía histórica y ecológica. In: Morrone JJ, Llorente-Bousquets J (eds) Una perspectiva latinoamericana de la biogeografía. UNAM, Mexico City, pp 141–154
Ruggiero A, Lawton JH, Blackburn TM (1998) The geographic ranges of mammalian species in South America: spatial patterns in environmental resistance and anisotropy. J Biogeogr 25:1093–1103
Schneider CJ, Smith TB, Larison B, Moritz C (1999) A test of alternative models of diversification in tropical rainforests: ecological gradients vs. rainforest refugia. Proc Natl Acad Sci U S A 96:13869–13873
Sclater PL (1858) On the general geographic distribution of the members of the class Aves. Proc Linn Soc London, Zool 2:130–145
Sclater WL, Sclater PL (1899) The geography of mammals. Kegan, Paul, Trench and Trübner, London
Simpson GG (1940) Mammals and land bridges. J Washington Acad Sci 30:137–163
Simpson GG (1965) The geography of evolution. Chilton, Philadelphia
Simpson GG (1977) Too many lines: the limits of the Oriental and Australian zoogeographic regions. Proc Am Philos Soc 121:107–120
Strayer DL, Power ME, Fagan WF, Pickett STA, Belnap J (2003) A classification of ecological boundaries. Bioscience 53:723–729
Takhtajan A (1986) Floristic regions of the world. University of California Press, Berkeley
Urtubey E, Stuessy TF, Tremetsberger K, Morrone JJ (2010) The South American biogeographic transition zone: an analysis from Asteraceae. Taxon 59:505–509
Vallejo B (2011) The Philippines in Wallacea. In: Telnov D (ed) Biodiversity, biogeography and nature conservation in Wallacea and New Guinea, vol I. The Entomological Society of Latvia, Riga, pp 27–42
van Oosterzee P (1997) Where worlds collide: the Wallace line. Cornell University Press, Ithaca
Verboom GA, Archibald JK, Bakker FT, Bellstedt DU, Conrad F, Dreyer LL, Forest F, Galley C, Goldblatt P, Henning JF, Mummenhoff K, Linder HP, Muasya AM, Oberlander KC, Savolainen V, Snijman DA, van der Niet T, Nowell TL (2009) Origin and diversification of the Greater Cape flora: Ancient species repository, hot-bed of recent radiation, or both? Mol Phylog Evol 51:44–53
Vilhena DA, Antonelli A (2015) A network approach for identifying and delimiting biogeographic regions. Nat Commun 6:1–9
Wallace AR (1860) On the zoological geography of the Malay Archipelago. J Proc Linn Soc 4:172–184
Wallace AR (1863) On the physical geography of the Malay archipelago. J R Geogr Soc 33:217–234
Wallace AR (1876) The geographical distribution of animals. Macmillan, London
Williams PH (1996) Mapping variations in the strength and breadth of biogeographic transition zones using species turnover. Proc R Soc B 263:579–588
Williams PH, de Klerk HM, Crowe TM (1999) Interpreting biogeographic boundaries among Afrotropical birds: spatial patterns in richness gradients and species replacement. J Biogeogr 26:459–474
Yarrow MM, Marín VH (2007) Toward conceptual cohesiveness: a historical analysis of the theory and utility of ecological boundaries and transition zones. Ecosystems 10:462–476
Zunino M (2005) Corotipos y biogeografía sistemática en el Euromediterráneo. In: Llorente Bousquets J, Morrone JJ (eds) Regionalización biogeográfica en Iberoamérica y tópicos afines: Primeras Jornadas Biogeográficas de la Red Iberoamericana de Biogeografía y Entomología Sistemática (RIBES XII.I-CYTED), Las Prensas de Ciencias. UNAM, Mexico City, pp 181–187

Chapter 2
What Is Evolutionary Biogeography?

Though this be madness, yet there is method in 't.

Hamlet, act II, scene III

Abstract Most of the authors involved in the theoretical development of evolutionary biogeography assume that dispersalism, panbiogeography, cladistic biogeography, and phylogeography represent alternative approaches. Instead, I consider that different biogeographic methods may be used to answer different questions, which are different steps of an integrative biogeographic analysis. This stepwise approach comprises five steps, each corresponding to particular questions and methods. Track analysis and methods for identifying areas of endemism are used initially to identify biotas (graphically represented on maps as generalized tracks or areas of endemism), which represent hypotheses of primary biogeographic homology and are the basic units of evolutionary biogeography. Then, cladistic biogeography uses available phylogenetic data to test the historical relationships between these biotas (secondary biogeographic homology). Based on the results of these analyses, a biogeographic regionalization is achieved. Intraspecific phylogeography, molecular dating, and fossils are incorporated to help identify the different cenocrons (set of taxa that share the same biogeographic history, which constitute identifiable subsets within a biota by their common biotic origin and evolutionary history) that became assembled in a biota. Finally, the geological and biological knowledge available is integrated to construct a geobiotic scenario that helps explain the way different dispersal and vicariance events contributed to biotic assembly and how the cenocrons dispersed to the biota analyzed. I present the concepts implied in these steps and some of the methods that may be applied to answer particular biogeographic questions and discuss how they can be integrated to explain biotic assembly within an integrative framework.

J. J. Morrone, *The Mexican Transition Zone*, https://doi.org/10.1007/978-3-030-47917-6_2

2.1 Introduction

As a consequence of alternating episodes of dispersal and vicariance, biotic evolution is rarely divergent, resulting in a reticulate rather than a branching structure (Upchurch and Hunn 2002; Riddle and Hafner 2004, 2006; Brooks 2005; Morrone 2009). To analyze these complex biogeographic patterns, we should try to discover the instances of vicariance as well as those of biotic convergence due to dispersal. (Some authors refer specifically to the different models of dispersal, namely, jump dispersal, diffusion, secular migration, and geodispersal (Morrone 2009), whereas others prefer to distinguish clearly dispersal for the aleatory overcoming of a barrier and dispersion for the gradual expansion of the distributional area of a species. I prefer to use the term dispersal in its more general meaning.) Currently, there are three general models in evolutionary biogeography (Morrone 2015): center of origin-dispersal-adaptation (CODA), vicariance, and dispersal-vicariance. CODA as originally formulated by Darwin (1859) and Wallace (1876) assumes a restricted origin of the ancestor of a group, followed by dispersal, arrival to new areas, and adaptation to new conditions. The vicariance model (Croizat 1958, 1964; Nelson and Platnick 1981) assumes widespread ancestors, which differentiate due to the appearance of barriers that isolate their populations. The dispersal-vicariance model (Reig 1981; Savage 1982) contemplates alternating episodes of dispersal and vicariance.

Both CODA and vicariance represent extreme, ideal situations, and I think that it is unrealistic to choose one of them and discard the other (Morrone 2009). The dispersal-vicariance model represents an integrative approach, already suggested by several authors (e.g., Reig 1981; Savage 1982; Brooks 2004; Lieberman 2004; Sanmartín and Ronquist 2004; Riddle et al. 2008; Crisci and Katinas 2009; de Queiroz 2014). Some decades ago, Cracraft (1975, p. 237) postulated that "when analyzing the history of biotas we must first attempt to understand the general patterns of vicariance, and then, following this, consider whether it is necessary to invoke dispersal to explain the composition of the biota." A few years later, George Gaylord Simpson considered that "A reasonable biogeographer is neither a vicarist nor a dispersalist but an eclecticist" (Simpson 1980, p. 253). Croizat added: "I do agree, but with the understanding that a biogeographer must be a vicarist in principle and a dispersalist in detail, case by case according to the merits of each case" (Croizat 1982, p. 297).

A stepwise approach (Morrone 2009) may allow to identify particular questions, to choose the most appropriate methods to answer them, and finally to integrate them in a coherent theory that explains biotic assembly (Fig. 2.1). Each step of this integrative approach corresponds to particular concepts, questions, and methods. It does not imply that every biogeographer must follow all the steps but that anybody may articulate a specific biogeographic question and choose the most appropriate method to answer it. Given some time, as the results of different biogeographic analyses accumulate, a coherent theory may be formulated by integrating them. I defend this approach within the philosophical framework of integrative pluralism,

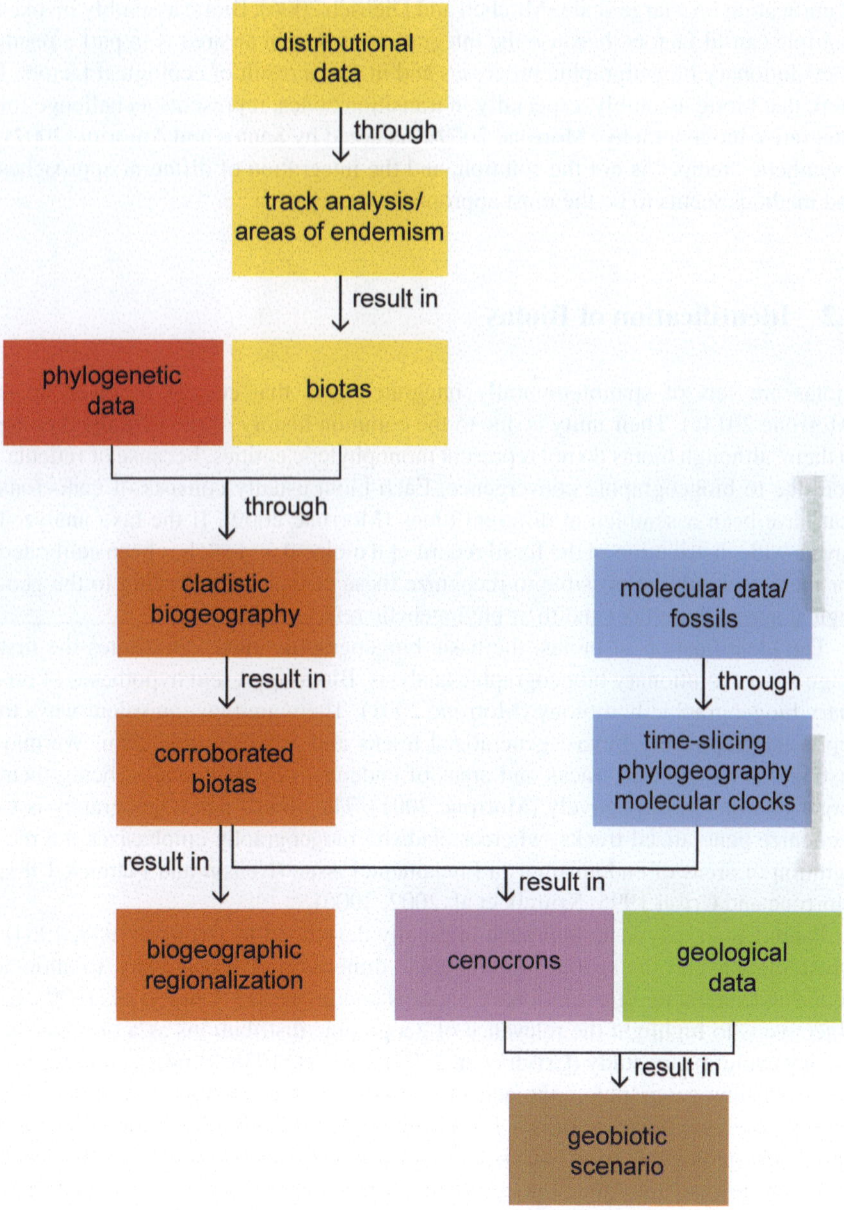

Fig. 2.1 Flow chart with the steps of an evolutionary biogeographic analysis

as conceived by Mitchell (2002). Integrative pluralism does not imply an eclectic or "anything goes" approach, but that different methods may be compatible because they give partial solutions, when answering particular questions. Integrative pluralism allows for the integration to explain a complex phenomenon without the need of unification on a large scale (Mitchell and Dietrich 2006). Biotic assembly invokes multiple causal factors, because the integration of taxa in an area is in part a result of evolutionary biogeographic processes and in part a result of ecological factors. I think that biotic assembly, especially in transition zones, represents a challenge for integrative biogeography (Morrone 2009). As noted by Santos and Amorim (2007), a synthetic "recipe" is not the solution, and the integration of different approaches and methods seems to be the most appropriate strategy.

2.2 Identification of Biotas

Biotas are sets of spatiotemporally integrated taxa that coexist in given areas (Morrone 2014a). Their unity is due to the common history of the taxa that belong to them, although biotas do not represent monophyletic entities, because of reticulation due to biogeographic convergence. Each biota usually consists of cenocrons that have been assembled at different times (Morrone 2009). If the taxa analyzed have a wide distribution in the fossil record or a molecular clock has been calibrated for them, it would be possible to recognize these cenocrons according to the geological age of these taxa and their phylogenetic relationships.

The identification of biotas, the basic biogeographic units, constitutes the first stage of an evolutionary biogeographic analysis. Biotas represent hypotheses of primary biogeographic homology (Morrone 2001). There are two equivalent ways to represent graphically biotas: generalized tracks and areas of endemism. We may distinguish generalized tracks and areas of endemism by their scales, being them larger or smaller, respectively (Morrone 2001). The aim of panbiogeography is to recognize generalized tracks, whereas cladistic biogeography emphasizes the recognition of areas of endemism as a fundamental issue (Nelson and Platnick 1981; Morrone and Crisci 1995; Szumik et al. 2002, 2006).

Panbiogeography is an approach originally developed by Croizat (1958, 1964), which emphasizes the spatial or geographic dimension of biodiversity, to allow a better understanding of evolutionary patterns and processes (Craw et al. 1999). Its objective is to highlight the relevance of geographic distributions as a prerequisite for any evolutionary study (Crisci et al. 2003). Croizat (1958, 1964) used to formulate the panbiogeographic approach in terms of three metaphors: "Earth and life evolve together," "space + time + form = the biological synthesis," and "life is the uppermost geological layer." Croizat intended to establish an independent science free from prior commitments to geological/geophysical theories (Craw and Page 1988). Panbiogeography is based on four assumptions (Craw et al. 1999): distributional patterns constitute an empirical database for biogeographic analyses; distributional patterns provide information about where, when, and how living organisms evolved; the spatial component of these distributional patterns can be represented graphically as generalized tracks; and testable hypotheses about historical relationships between biotic distributions and earth history are derived from correlating these distributional patterns with geological/geomorphological features. Although

Croizat's metaphors are useful for understanding broad, general patterns, biotic assembly is a more complex issue, involving episodes of both dispersal and vicariance (Morrone 2015). Thus, track analysis may be applied to identify biotas, constituting the first step of an evolutionary biogeographic analysis.

A track analysis (Morrone 2009, 2015) comprises three successive steps (Fig. 2.2):

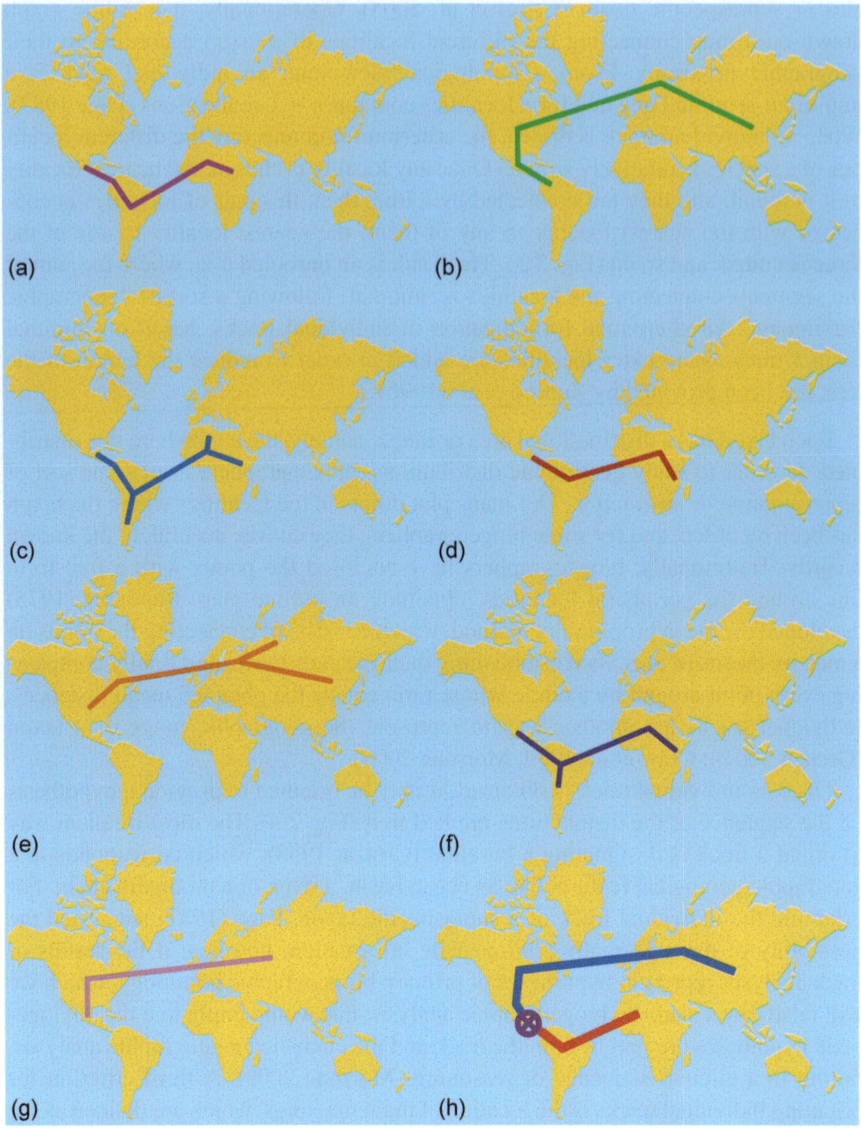

Fig. 2.2 Steps of a track analysis. (**a**–**g**) Obtaining individual tracks; (**h**) identifying generalized tracks and node

1. Constructing individual tracks for two or more different taxa, by connecting the localities of each taxon according to their geographical proximity
2. Obtaining generalized tracks based on the superposition of two or more individual tracks
3. Identifying nodes in the areas where two or more generalized tracks intersect

Individual Tracks An individual track represents the spatial coordinates of a species or supraspecific taxon (Crisci et al. 2003). Operationally, it is a line graph drawn on a map connecting the different localities of a taxon according to their geographic proximity. From a topological viewpoint, an individual track is a minimum-spanning tree that for n localities contains $n - 1$ connections (Page 1987). When an individual track is drawn, the criterion for connecting the different localities of a species is relatively simple. Once any locality is chosen, the nearest locality to it is found, and they are connected by a line; then, this pair of localities is connected with the nearest locality to any of them; the nearest locality to any of the three is united, and so on (Fig. 2.3). The result is an unrooted tree, where the sum of the segments connecting the localities is minimal, following a sort of "geographic parsimony." An alternative formalization of individual tracks, based on minimal Steiner trees where extra localities are added in order to reduce the length of the tree, has been provided by Zunino et al. (1996).

Each taxon has a distributional area or range, namely, the area where it is distributed. In order to study geographic distributions, biogeographers need some sort of representation or abstraction. Dot maps plot points in the localities where the taxon has been recorded, and for some biogeographers, they convey accurately the known records. Traditionally, biogeographers have enclosed the points with a free-form line around the peripheral localities, obtaining an outline map. Rapoport (1975) developed the mean propinquity method, which consists of connecting the points on a map by means of arcs, then establishing their mean distance, and finally compassing every point around by a circle whose ratio equals the obtained mean distances. Individual tracks are another way to represent the geographic range of a taxon (Grehan 2001b; Craw et al. 1999; Morrone 2009).

Once an individual track is obtained, it may be oriented to provide a hypothesis on the sequence of the disjunctions implied in it (Fig. 2.4). The most frequent way to orient a track is designating a baseline (Croizat 1958), which corresponds to a geographic/geological feature, like an ocean basin, a river, or a mountain chain. For orienting the individual track of a supraspecific taxon, Page (1987) suggested the possibility of using formally phylogenetic information; however, if the results of track analyses represent hypotheses of primary biogeographic homology, which we will falsify by a cladistic biogeographic analysis, this would imply that the phylogenetic hypotheses are part of both the track and the cladistic biogeographic analyses, falling in a circular sequence of reasoning (Morrone 2009). A third criterion for orienting individual tracks is the location of main massings, which are defined as the greatest concentration of biological diversity within the range of the taxon, e.g., number of species or genetic diversity. In general, main massings represent areas of

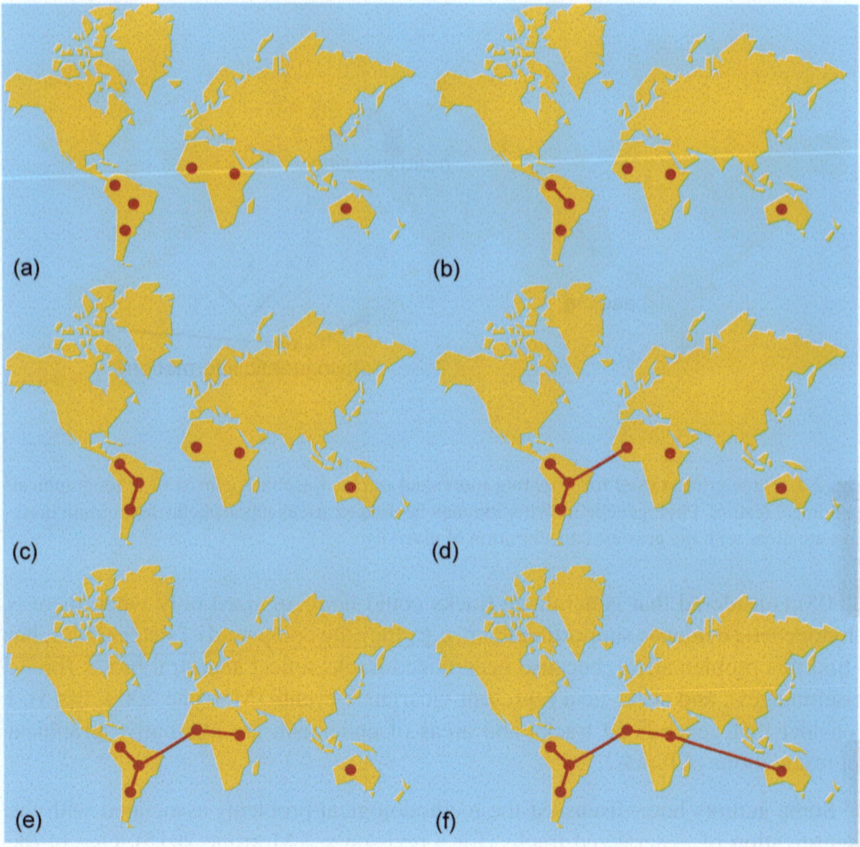

Fig. 2.3 Obtaining an individual track. (**a**) Localities of distribution of a species; (**b**) choose a locality and join it to its nearest locality; (**c–f**) joining the remaining localities based on their proximity

numerical, genetic, or morphological diversity of a group (Page 1987), but if a track is oriented from the main massing toward the periphery, the inference involved would be similar to that from dispersal biogeographers (Crisci et al. 2003), so this criterion might be also inappropriate. Of the three criteria for orienting tracks, the less problematic is the baseline, but when the analyses are undertaken on continental scale, the use of geological or tectonic characteristics is somewhat more difficult to carry out (Morrone 2004a). For these reason, published track analyses do not generally orient the individual tracks (Morrone 2009, 2015).

Generalized Tracks Generalized or standard tracks result from the significant superposition of different individual tracks (Zunino and Zullini 1995). They indicate the preexistence of ancestral biotas, which became fragmented by geological, tectonic, or ecological events (Craw 1988). Generalized tracks are obtained comparing the individual tracks and looking for a significant agreement. Nihei and Carvalho

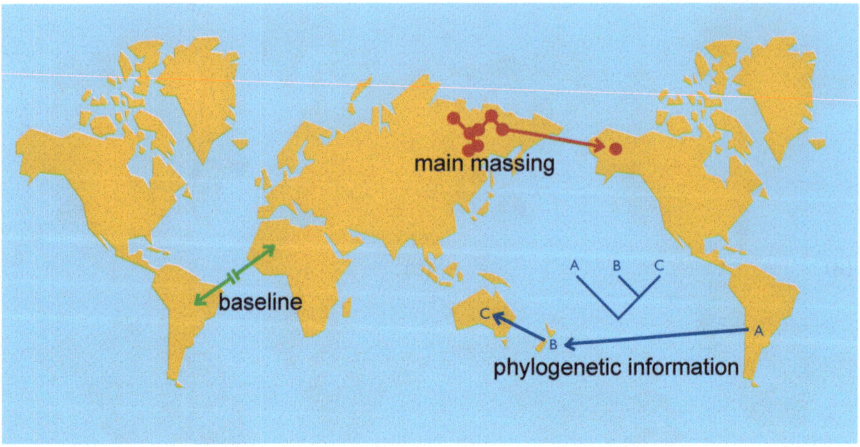

Fig. 2.4 Three criteria used for orienting individual tracks. Baselines consist of a geographical/geological feature. Phylogenetic information may be used to orient supraspecific taxa. Main massings are areas with the greatest concentration of diversity

(2005) considered that generalized tracks could be recognized only when there is phylogenetic evidence supporting them, e.g., they are comprised of sister clades, but I find this problematical, because generalized tracks reflect ancestral biotas (biotic assemblages), and sister taxa represent vicariance events (Morrone 2009, 2015). I consider that generalized tracks and areas of endemism are alternative graphical representations of biotas.

Some authors have discussed the methodological problems associated with the identification of generalized tracks (for a revision see Morrone 2015). One of the most important is the arbitrary decision on how good should be the congruence among the individual tracks to be considered part of a generalized track. Ferrari et al. (2013) evaluated empirically this issue, by comparing the results of three quantitative methods: geometric distance between segments of individual tracks as implemented in program MartiTracks (Echeverría-Londoño and Miranda-Esquivel 2011), track compatibility using program CLIQUE of PHYLIP (Felsenstein 1986), and parsimony analysis of endemicity with progressive character elimination using the phylogenetic software TNT (Goloboff et al. 2008). They found that none of these approaches solved the congruence problem objectively, although parsimony analysis of endemicity provided the most reliable results.

Nodes Nodes are complex areas, where two or more generalized tracks intersect, which are usually interpreted as tectonic and biotic convergence zones (Heads 2004). The recognition of nodes represents one of the most important contributions of panbiogeography, because they allow us to speculate on the existence of transition zones (Morrone 2009; Miguel-Talonia and Escalante 2013). Nodes may represent the location of endemism, high diversity, distributional boundaries, disjunction, "anomalous" absence of taxa, incongruence and convergence of characters, and unusual hybrids, among other features (Heads 2004). Fontenla and López Admirall (2008) considered that endemism might not be a relevant feature of nodes, because rather than exclusive species they are based on species from different generalized tracks. Miguel-Talonia and Escalante (2013) suggested that the characteristics listed by Heads (2004) depend on the scale and the taxa analyzed. Escalante et al. (2017) considered that areas with several nodes, named "node-diverse," probably represent transition areas with multiple biotic histories superimposed.

In order to provide an objective procedure to identify nodes, Henderson (1989) suggested that they may correspond with points with a higher density of terminal track vertices, like 1° vertices. These are endpoint vertices, found at the periphery of a generalized track, which only have one connecting link to another point. Miguel-Talonia and Escalante (2013) suggested that there nodes corresponding to 1° vertices are more relevant in evolutionary terms, whereas those corresponding to 2° or more vertices may be related to ecological processes. In order to represent nodes graphically on a map, Fortino and Morrone (1997) suggested to use an "x" enclosed by a circle, although this representation does not represent a geographic surface or has a precise location or ecological/geographic characteristics (Escalante et al. 2017).

Areas of Endemism Areas of endemism are defined as geographic regions comprising the distributions of two or more monophyletic taxa that exhibit phylogenetic and distributional congruence and having their respective relatives occurring in other such-defined regions (Harold and Mooi 1994) or as areas of nonrandom distributional congruence among different species or supraspecific taxa (Morrone 1994). Congruence does not demand complete agreement on those limits at all possible scales of mapping (Wiley 1981; Morrone 1994; Hausdorf 2002; Apodaca and Crisci 2018). Both historical and ecological factors are invoked when explaining endemism: historical events (usually vicariance) explain how taxa are confined to the areas of endemism, whereas ecological explanations (biotic and abiotic factors) deal with their present limits (Morrone 2008). Crother and Murray (2011; see also Murray and Crother 2016) have considered that areas of endemism ontologically are individuals, that change over geologic history and evolutionary time, due to species expansions and contractions, speciation, and extinction.

In order to identify areas of endemism, Müller's (1973) protocol for working out "dispersal centers" has been applied (Morrone et al. 1994; Roig-Juñent 1994). This protocol consists in plotting the ranges of species on a map and finding the areas of congruence between several species, assuming that the species distributions are relatively small compared with the region itself, that the limits of these distributions are known with certainty, and that the validity of the species is not in dispute. According to Linder (2001), areas of endemism must have at least two endemic species, the distributions of the species endemic to them should be maximally congruent, they should be narrower than the whole study area, and they should be mutually exclusive.

The recognition of areas of endemism is usually based on distributional data, without considering the divergence times of the species analyzed. When a temporal dimension is incorporated to the analysis of endemism, there are four alternative scenarios that have been proposed to explain the assembly of taxa (Noguera-Urbano 2016):

Area of endemism composed of asynchronous taxa (different temporal strata): the area of endemism is supported by at least two related or unrelated taxa with different divergence times.

Area of endemism with synchronous taxa (one temporal stratus): the area of endemism is defined by at least two taxa that have a similar divergence time.

Area of endemism with related synchronous taxa and unrelated synchronous taxa: the area of endemism is structured by at least four taxa having similar divergence time between pairs, such that they all form a single temporal stratus.

Area of endemism composed of one or more synchronous congruent endemic taxa and one or more asynchronous congruent endemic taxa: the area of endemism is integrated by at least three taxa, which may be sister taxa or not, having different multiple temporal strata.

There are some problems concerning the identification of areas of endemism. Crisp et al. (1995) considered that the alternative procedures for identifying areas of endemism were controversial, specially questioning whether the hierarchical model of parsimony analysis of endemicity (PAE) was adequate for that purpose. Humphries and Parenti (1999) argued that including species that are ecologically very different can help argue for a historical, rather than ecological explanation for the areas of endemism identified. Linder (2001) proposed three optimality criteria to help choose the best estimate of the areas of endemism: the number of areas identified, the proportion of the species restricted to the areas of endemism, and the congruence of the distributions of the species restricted to the areas of endemism. Roig-Juñent et al. (2002) enumerated some problems with the identification of areas of endemism, namely, lack of distributional data, bias toward locality data, and subjectivity when drawing the exact limits of the areas of endemism. DaSilva et al. (2015) considered that the definition and criteria used for the identification of areas of endemism have not been sufficiently discussed in the literature.

Methods There are several methods that can be applied in track analysis and the identification of areas of endemism (Szumik et al. 2002; Crisci et al. 2003; Morrone 2004a, 2009; Noguera-Urbano 2016). I will deal herein with the most commonly applied method, parsimony analysis of endemicity (Rosen 1988; Morrone 1994, 2014b).

Parsimony analysis of endemicity (PAE) was formulated originally by Rosen (1985) and fully developed by Rosen (1988) and Rosen and Smith (1988). It is also known as parsimony analysis of shared presences (Rosen and Smith 1988), parsimony analysis of distributions (Trejo-Torres and Ackerman 2001), parsimony analysis of species sets (Trejo-Torres 2003), cladistic analysis of distributions and endemism (Porzecanski and Cracraft 2005), and parsimony analysis of community assemblages (Ribichich 2005). Parsimony analysis of endemicity constructs cladograms based on the parsimony analysis of presence-absence data matrices of species and supraspecific taxa (Cracraft 1991; Myers 1991; Morrone 1994, 2014b; Escalante et al. 2003). PAE cladograms allow different biogeographic processes: synapomorphies are interpreted as vicariance events, parallelisms as dispersal events, and reversals as extinction events.

Crisci et al. (2003) distinguished three varieties of PAE, according to the geographical units analyzed, namely, localities, areas of endemism, and grid cells. There are other units that have been used in parsimony analyses of endemicity, namely, hydrological basins (Aguilar-Aguilar et al. 2003), real and virtual islands (Maldonado and Uriz 1995; Trejo-Torres and Ackerman 2001), transects (Trejo-Torres and Ackerman 2002; García-Trejo and Navarro 2004; León-Paniagua et al. 2004; Navarro et al. 2004), communities (Ribichich 2005), and political entities (Cué-Bär et al. 2006). García-Barros (2003) proposed an alternative classification, based on the objectives of the analysis: to infer historical relationships between areas, to identify areas of endemism, and to classify areas. PAE may be used for track analyses, where the clades obtained are mapped as generalized tracks (Craw et al. 1999; Luna-Vega et al. 2000; Morrone and Márquez 2001). Luna-Vega et al. (2000) and García-Barros et al. (2002) have proposed that once the most parsimonious cladograms have been obtained, it is possible to remove or exclude the taxa supporting the different clades and analyze the reduced matrix to search for alternative clades supported by other taxa. This procedure has been named parsimony analysis of endemicity with progressive character elimination or PAE-PCE (Echeverry and Morrone 2010). In order to root the PAE cladogram(s), a hypothetical area with all "0" is added to the matrix. Cano and Gurrea (2003) and Ribichich (2005) have used an area coded with all "1," thus grouping areas according to shared absences, which would imply depletion through time starting from a cosmopolitan biota (Cecca 2002).

Parsimony analysis of endemicity has received several criticisms (for a revision see Morrone 2014b). Linder and Mann (1998) criticized Morrone's (1994) approach for identifying areas of endemism with PAE, because grid cells can only be used as presence-absence data, and under-collecting may result in grid cells being omitted. Some authors considered that PAE is not a valid historical method, because it does not take into account the phylogenetic relationships of the taxa analyzed (Humphries 1989, 2000; García-Barros et al. 2002; Santos 2005). According to Rosen (1988) and Nihei (2006), there are two possible interpretations for PAE cladograms: static and dynamic. The former assumes that cladograms constitute an alternative to phenetic classification methods, whereas according to the latter, cladograms are hypotheses on the historical or ecological relationships of the areas analyzed. If we interpret the external area with all "0" as an area lacking suitable conditions for the taxa to survive therein (ecological interpretation), relationships will indicate ecological affinities. If we interpret the external area as a geologically ancient area, where none of the taxa has yet evolved (historical interpretation), relationships will indicate biotic dispersal or vicariance events. Most of the authors that have used PAE explored historical interpretations of the detected patterns, usually from a vicariance viewpoint; for ecological interpretations, see Trejo-Torres and Ackerman (2002), Trejo-Torres (2003) and Ribichich (2005). Nihei (2006) presented a revision of PAE, including a discussion of its history and applications. He considered that most of the criticisms dealt with its methodology rather than on its theory and that they usually resulted from the confusion between the dynamic and static approaches. Nihei (2006) warned biogeographers applying PAE to be aware of the problems and limitations of both dynamic and static PAE and evaluate new variations of PAE. Morrone (2014b) provided a review of the critiques to PAE and its use as a method of evolutionary biogeography. Escalante (2015) compared parsimony analysis of endemicity (Rosen 1988; Morrone 1994) and endemicity analysis (Szumik et al. 2002, 2006), applied to identifying areas of endemism, and found that the former performed better in terms of strict sympatry.

The algorithm of parsimony analysis of endemicity with progressive character elimination or PAE-PCE (Echeverry and Morrone 2010; Morrone 2014b) comprises the following steps (Fig. 2.5):

1. Choose a set of biogeographical units across the study area, for example, localities (Fig. 2.5a), pre-defined areas of endemism, or geographic areas defined by physiographical criteria (Fig. 2.5b) or grid cells (Fig. 2.5c).
2. Determine the geographical distribution of the taxa being analyzed, by simply recording their localities (Fig. 2.5d), constructing individual tracks (Fig. 2.5e), or modeling their distributions using niche modeling (Fig. 2.5f). If available, consider adding phylogenetic information from supraspecific taxa (Fig. 2.5g).
3. Construct an r × c matrix (Fig. 2.5h), where r (rows) represents the biogeographical units analyzed and c (columns) represents the species and/or supraspecific taxa. Code each entry as either 1 or 0, depending on whether each taxon is present or absent in the unit. Add a hypothetical unit coded as all zeros to the matrix in order to root the resulting area cladogram(s).
4. Analyze the matrix with a parsimony algorithm. If more than one area cladogram (Fig. 2.5i) is found, calculate a strict consensus cladogram.
5. Identify biotas in the resulting cladogram as the monophyletic groups of units defined by at least two taxa (= geographic synapomorphies). Additionally, if an historical interpretation is being applied, infer specific biogeographical processes from the optimized taxa onto the cladogram: synapomorphies as vicariance events, parallelisms as dispersal events, and reversals as extinction events.
6. Represent the biotas identified in the previous step on a map as areas of endemism (groups of grid cells-Fig. 2.5j or coarse maps-Fig. 2.5k) or generalized tracks (Fig. 2.5l).

Criteria for Evaluating Areas of Endemism DaSilva et al. (2015) proposed a protocol to identify areas of endemism, combining quantitative and qualitative criteria. After quantitative analyses are undertaken using parsimony analysis of endemicity and endemicity analysis, six qualitative criteria (Fig. 2.6) are applied:

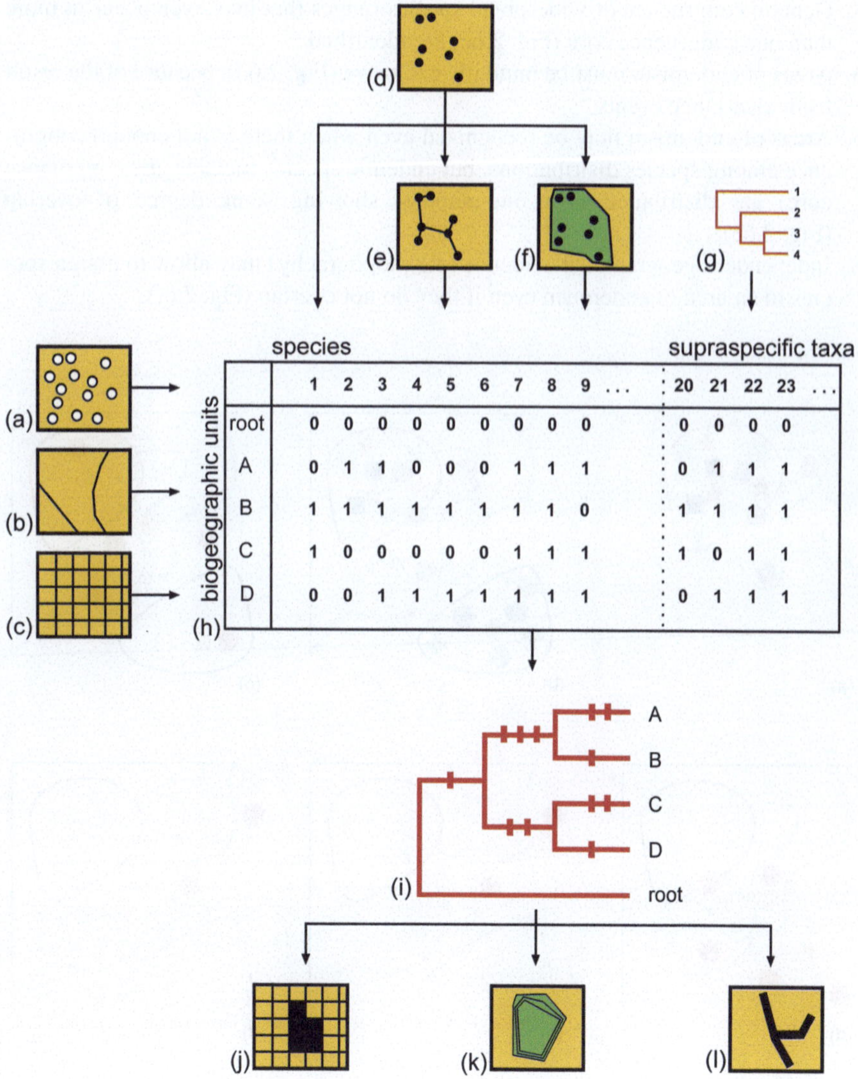

Fig. 2.5 Flowchart showing the steps of parsimony analysis of endemicity (PAE). (**a**) Localities analyzed; (**b**) pre-defined areas of endemism or areas defined by physiographical criteria; (**c**) grid cells; (**d**) locality records; (**e**) individual tracks; (**f**) modelled distributions; (**g**) phylogenetic information from supraspecific taxa; (**h**) data matrix; (**i**) area cladogram obtained; (**j**) areas of endemism as groups of grid cells; (**k**) areas of endemism as coarse maps; (**l**) generalized tracks

1. The congruence of at least two species showing considerable overlap (Fig. 2.6a) defines the "congruence core" of the area of endemism.
2. Partially overlapping species that extend outside the congruence core (Fig. 2.6b) define the "maximum region of endemism."

3. General congruence of widespread species ranges that may even occur in more than one congruence core (Fig. 2.6c) are identified.
4. Areas of endemism must be mutually exclusive (Fig. 2.6d), because of the result from vicariance events.
5. Areas of endemism may be recognized even when there is not enough congruence among species distributions, but endemic species (outside any congruence core) are distributed near one another showing some degree of overlap (Fig. 2.6e).
6. Independent geographical evidence (e.g., topography) may allow to assign species to an area of endemism even if they do not overlap (Fig. 2.6f).

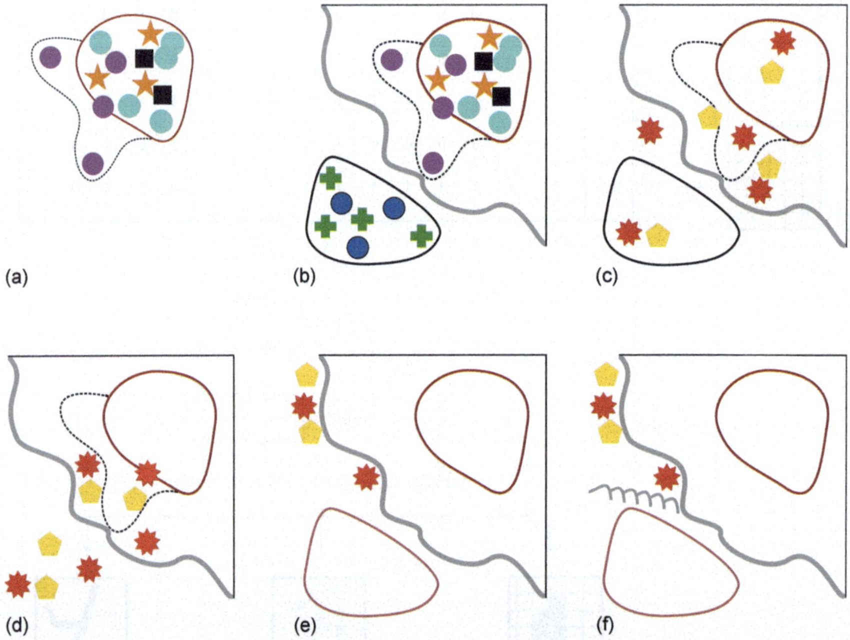

Fig. 2.6 Hypothethical examples of the use of six qualitative combined criteria (C1–C6) to identify areas of endemism. (**a**) C1: delimitation of a congruence core (CC, solid line) based on three species (green circle, orange star, and black square); (**b**) C2: delimitation of a maximum region of endemism (dashed line) based on a species occurring in a CC but not occurring in any other (purple circle); (**c**) C3: avoiding the delimitation of an area of endemism by congruence range of widespread species (red sun and yellow pentagon); (**d**) C4: avoiding overlap of CCs, as they must be mutually exclusive (one of them could be delimited by white sun and white pentagon species); (**e**) C5: two endemic species (red sun and yellow pentagon) may be evidence of another area of endemism, even with poor range congruence; (**f**) C6: corroborating the new area of endemism shown in (**e**), because it is on the other side of a large river and a mountain range (gray lines) of the same topographical unit (modified from DaSilva et al. 2015)

2.3 Testing Relationships Among Biotas

Since generalized tracks are unrooted, they connect geographic areas but do not specify a precise sequence of fragmentation. For example, given a generalized track joining the Sierra Madre Occidental, the Sierra Madre Oriental, and the Chiapas Highlands, which of the three areas separated first from the others? In order to determine this sequence, phylogenetic data need to be incorporated. Cladistic biogeography assumes that there is a correspondence between the phylogenetic relationships of the taxa and the relationships between the areas that they inhabit (Platnick and Nelson 1978; Nelson and Platnick 1981). Cladistic biogeography uses information on the cladistic relationships between the taxa and their geographic distribution to postulate hypotheses on relationships between areas. If several taxa show the same pattern, such congruence is evidence of a common history (Wiley 1988a; Morrone and Crisci 1995; Zunino and Zullini 1995; Enghoff 1996; Humphries and Parenti 1999). We may characterize cladistic biogeography considering that it originated from the joining of three independent research programs: Hennig's (1950) phylogenetic systematics, Croizat's (1958, 1964) panbiogeography, and Wegener's (1929) continental drift. To them, Nelson and Platnick (1981) added the deductive-hypothetical method of Popper (1959, 1963), although there is no consensus regarding the falsifiability of cladistic biogeographic hypotheses in a Popperian sense (see Santos and Capellari 2009).

It is possible to raise an analogy between systematics and biogeography (Morrone 2009). In systematics, we study taxa, and we classify them by their shared characters, whereas in biogeography, we study areas, classifying them by their shared taxa. This implies an equivalence between taxa (systematics) and areas (biogeography). This correspondence, however, has been put in doubt by Hovenkamp (1997, 2001), who suggested that instead of reconstructing the sequence of area fragmentation, we should analyze the sequence of vicariance events. According to Nelson and Platnick (1978), cladistic biogeography poses three questions: is endemism geographically nonrandom and, if so, which areas of endemism can be identified?; given some areas of endemism, are the interrelationships of their endemic taxa geographically nonrandom and, if so, what is the pattern formed by their interrelationships?; and given one or more patterns of interrelationships, as represented by one or more general area cladograms, does the pattern correlate with the geological history?

Cladistic biogeography is based on geographic congruence, e.g., the finding of identical patterns between unrelated taxa is interpreted as having a common cause. For example, the breakup of the supercontinent Pangaea 250 m.y.a. produced a general pattern of vicariance between different groups of continental taxa, or the uplift of the Isthmus of Panama produced general patterns of vicariance in different groups of marine organisms. Congruence is detected once an initial pattern has been established. In cladistic biogeography, secondary biogeographic homology (Morrone 2001) is usually considered to be the result of vicariance, although there may be instances of congruence due to geodispersal (Lieberman 2000). This is why this approach, although originally known as "vicariance biogeography" (e.g., Rosen and Nelson 1980; Nelson and Platnick 1981), is now widely known as "cladistic biogeography" (Parenti 1981, 2007; Page 1988; Humphries and Parenti 1999).

A cladistic biogeographic analysis comprises three basic steps (Fig. 2.7):

1. Constructing taxon-area cladograms, from the taxonomic cladograms of two or more different taxa, by replacing their terminal taxa with the areas they inhabit
2. Obtaining resolved area cladograms from the taxon-area cladograms (when demanded by the method applied)
3. Obtaining a general area cladogram, based on the information contained in the resolved area cladograms

Taxon-Area Cladograms Taxon-area cladograms are obtained by replacing the name of each terminal taxon in the cladograms of the taxa analyzed, by the area where it is distributed. For example, if a taxon (1 (2 (3, 4))) has species 1 distributed in North America, species 2 in Central America, species 3 in South America, and species 4 in the Antilles, by replacing the four species in the cladogram by the areas where they are distributed, we obtain the following taxon-area cladogram: (North America (Central America (South America, Antilles))).

Resolved Area Cladograms The construction of taxon-area cladograms is simple when each taxon is endemic to a single area and each area has only one taxon, but it is more complex when taxonomic cladograms include widespread taxa, redundant distributions, and missing areas. In these cases, some methods require that taxon-area cladograms are turned into resolved area cladograms (Nelson 1984; Page 1988, 1993; Morrone and Carpenter 1994; Sanmartín and Ronquist 2002).

Widespread taxa occur when any of the terminal taxa of a taxon-area cladogram inhabits two or more of the studied areas (Nelson and Platnick 1981). They are also known as masts (for "multiple areas on a single terminal"; Ebach et al. 2005). For example, if a taxon (1 (2, 3)) has species 1 distributed in both North America and Central America, species 2 in South America and species 3 in the Antilles, by replacing the three species in the cladogram by the areas where they are distributed, we obtain the following taxon-area cladogram: (North America-Central America (South America, Antilles)). As a result of the widespread taxon, North America and Central America appear together in the taxon-area cladogram.

In order to resolve widespread taxa, three assumptions have been proposed (Morrone 2009). Under assumption 0 (Zandee and Roos 1987), the areas inhabited by a widespread taxon are considered as a monophyletic group in the resolved area cladogram, meaning that the taxon is treated as a synapomorphy of the areas. Under assumption 1 (Nelson and Platnick 1981), the widespread taxon is not considered as a synapomorphy when constructing the resolved area cladograms, and the areas inhabited by it can constitute mono or paraphyletic groups in the resolved area cladograms. Under assumption 2 (Nelson and Platnick 1981), only one occurrence is considered as evidence, whereas the other can "float" in the resolved area cladograms, therefore constituting the areas involved mono, para, or polyphyletic groups.

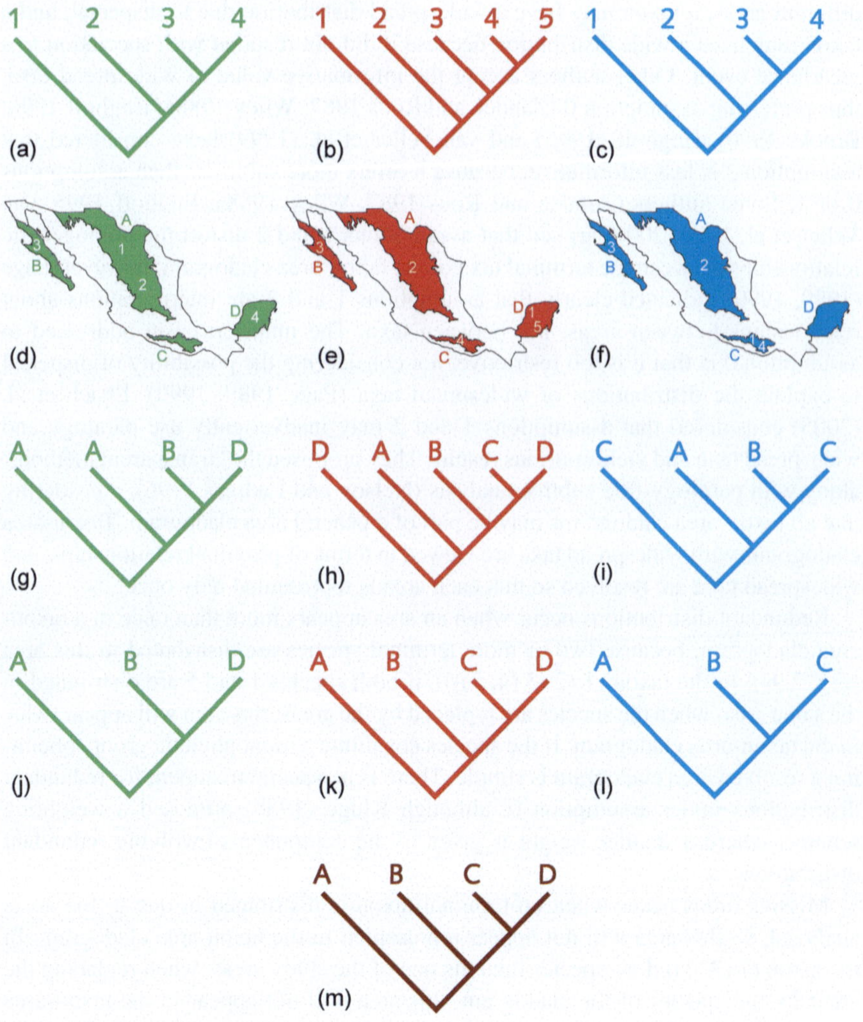

Fig. 2.7 Steps of a cladistic biogeographic analysis. (**a–c**) Taxonomic cladograms; (**d–f**) maps showing the distribution of the species of the three taxa analyzed; (**g–i**) taxon-area cladograms; (**j–l**) resolved area cladograms; (**m**) general area cladogram

The three assumptions show an inclusion relationship, since topologies obtained under assumption 0 are included within those obtained under assumption 1, and those obtained under assumption 1 are as well included within those from assumption 2 (Fig. 2.8).

Some authors (Nelson and Platnick 1981; Humphries 1989, 1992; Morrone and Carpenter 1994) prefer assumption 2, considering that widespread taxa are a source of ambiguity, because a future analysis can show that a widespread taxon really represents two or more different taxa, not necessarily related, and inhabiting different areas, a taxon may have a widespread distribution due to dispersal, and a taxon may have a wide distribution because it did not respond with speciation to a vicariance event. Other authors accept the informative value of widespread taxa, thus preferring assumption 0 (Zandee and Roos 1987; Wiley 1988a; Enghoff 1996; Brooks 1990). Enghoff (1995) and van Veller et al. (1999) have considered that assumption 2 is less informative, because it offers more solutions than assumptions 0 or 1. Some authors (Zandee and Roos 1987, Wiley 1988a, Enghoff 1996, van Veller et al. 1999, 2000) argued that assumptions 1 and 2 distort the phylogenetic relationships between the terminal taxa of the taxon-area cladogram; however, Page (1989, 1990) indicated clearly that assumptions 1 and 2 are interpretations about relationships between areas, not between taxa. The main criticism addressed to assumption 0 is that it is too restrictive, not considering the possibility of dispersal to explain the distributions of widespread taxa (Page 1989, 1990). Ebach et al. (2005) considered that assumptions 1 and 2 may inadvertently use paralogy and widespread taxa and yield spurious results. They proposed the "transparent method," along with paralogy-free subtree analysis (Nelson and Ladiges 1996), considering that all taxon-area cladograms may be part of a general area cladogram. Taxon-area cladograms with widespread taxa are viewed in terms of proximal relationships, and widespread taxa are resolved so that each area is represented only once.

Redundant distributions occur when an area appears more than once in a taxon-area cladogram, because two or more terminal species are distributed in this area (Fig. 2.9a). In the taxon (1 (2 (3 (4, 5)))), if both species 1 and 5 are distributed in the same area, when the species are replaced by the areas, this area will appear twice in the taxon-area cladogram. If the species constitute a monophyletic group, obtaining a resolved area cladogram is simple. There is no special treatment for redundant distributions under assumption 0, although Kluge (1988) proposed a weighting scheme, where a smaller weight is given to the components involving redundant distributions.

Missing areas occur when no terminal taxon is distributed in one of the areas analyzed, so this area will not appear represented in the taxon-area cladogram. In the taxon (1 (2, 3)) if no species inhabits one of the study areas, when replacing the areas by the species of the cladogram, this area will not appear in the taxon-area cladogram (Fig. 2.10a). Missing areas, which are caused by extinction or insufficient studies, are treated as non-informative, coding them with "?", so that they can be placed in all the possible positions in the resolved area cladograms (Fig. 2.10b–f). Also it is possible to treat them as primitively absent, coding them with "0" (Kluge 1988).

General Area Cladograms Based on the information from the different resolved area cladograms, a general area cladogram is derived. It represents a hypothesis on the biogeographic history of the taxa analyzed and the areas where they are distributed. The general area cladogram that results from the analysis may be falsified with a geological area cladogram, which is an area cladogram based on geological or tectonical data (Rosen 1985; Seberg 1991; Swenson et al. 2001; van Welzen et al. 2001). Another way to evaluate general area cladograms is calculating items of error

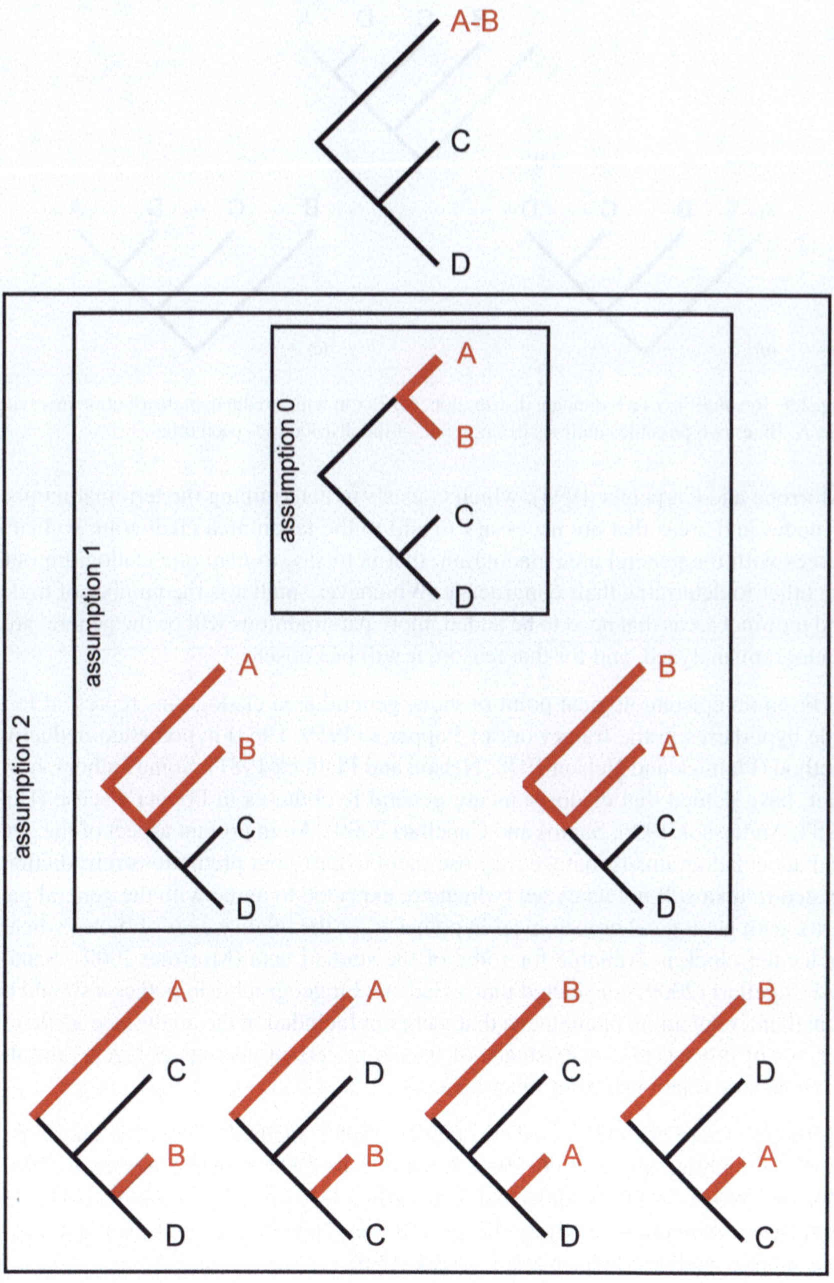

Fig. 2.8 Possible resolved area cladograms obtained for a taxon-area cladogram with a widespread taxon, as monophyletic (assumption 0), mono and paraphyletic (assumption 1), and mono, para, and polyphyletic groups of areas (assumption 2)

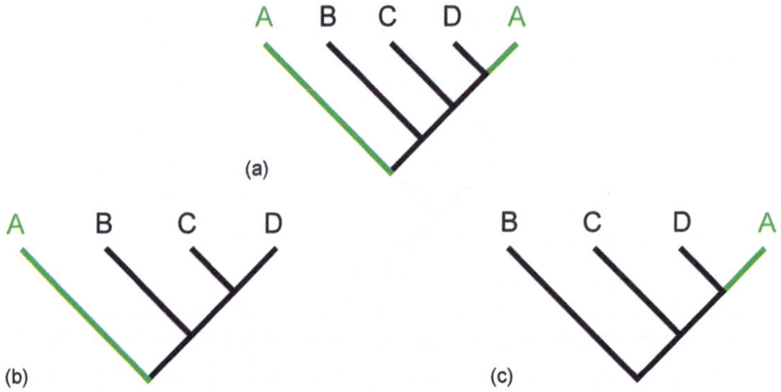

Fig. 2.9 Resolutions of a redundant distribution. (**a**) Taxon with a redundant distribution involving area A; (**b**, **c**) two possible solutions deleting one of the distributions each time

(Morrone and Carpenter 1994), which consists in determining the terminal number of nodes and areas that are necessary to add to the taxon-area cladogram so that it agrees with the general area cladogram, that is to say, to map one cladogram onto the other to determine their congruence. Whichever smaller is the number of nodes and terminal areas that need to be added, more parsimonious will be the general area cladogram analyzed, and for that reason, it will be chosen.

From an epistemological point of view, general area cladograms represent testable hypotheses in the framework of Popper's (1959, 1963) hypothetico-deductive method (Platnick and Nelson 1978; Nelson and Platnick 1981). Some authors, however, have denied that cladograms are general hypotheses in Popper's sense (Hull 1983; Andersson 1996; Santos and Capellari 2009). An important aspect of the general area cladograms is that we may use them to carry out predictions/retrodictions related to taxa still not analyzed (which are expected to agree with the general pattern), with geological or tectonical hypotheses, or the relative ages of biotas when a molecular clock is available for some of the studied taxa (Morrone 2009). Santos and Capellari (2009) considered that a cladistic biogeographic hypothesis should be consilient, explaining phenomena that were not included in the analysis, e.g., distributions of other taxa, the existence of fossils in certain geological layers, and the phylogenetic relationships of other taxa.

Methods There are many cladistic biogeographic methods (Morrone and Crisci 1995; Humphries and Parenti 1999; Crisci et al. 2003; Goyenechea et al. 2001; Morrone 2004a, 2009; Ronquist and Sanmartín 2011; Arias 2017). I will deal herein with Brooks parsimony analysis (Wiley 1987) and parsimony analysis of paralogy-free subtree analysis (Nelson and Ladiges 1996).

Brooks parsimony analysis (BPA) was proposed by Wiley (1987, 1988a, b) and is based on the ideas developed initially by Brooks (1981, 1985) for historical ecology. It is a parsimony analysis of taxon-area cladograms that are codified as two-state variables and analyzed as characters (Vargas 1992; Biondi 1998; van Veller

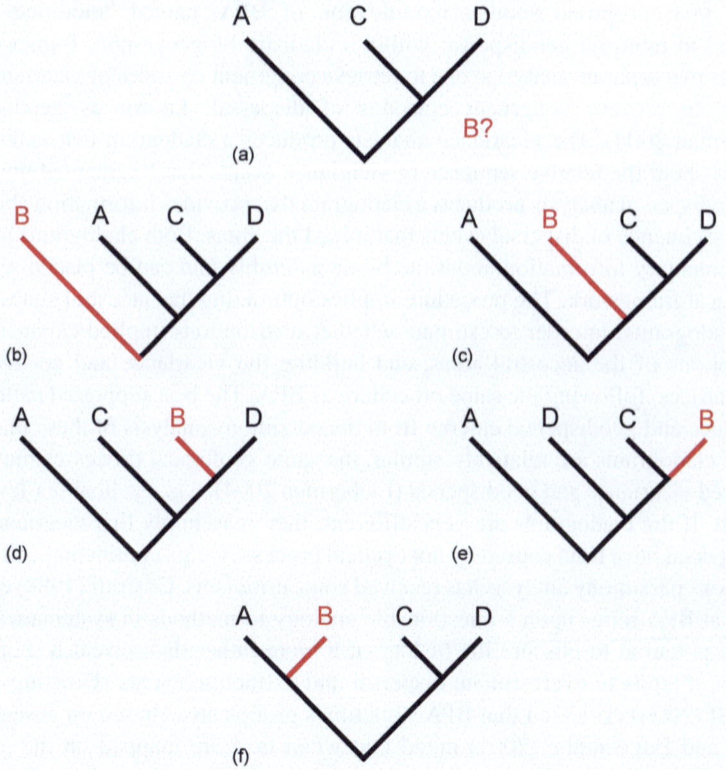

Fig. 2.10 Resolutions of a taxon with a missing area. (**a**) Taxon with a missing area; (**b–f**) five possible solutions placing it in all the possible positions in the cladogram

et al. 2000; Brooks 2004). In order to apply BPA, a data matrix is constructed based on the taxon-area cladograms, and it is analyzed with a parsimony algorithm.

An alternative implementation of BPA was proposed by Kluge (1988), which differs in three aspects. It considers that missing areas are uninformative, coding them with "0." It considers that widespread taxa, caused either by dispersal or by not having responded to vicariance, are irrelevant, and thus should not be taken into account. Since for redundant distributions it is impossible to determine which distribution is irrelevant (by being due to dispersal) and which one is not, Kluge (1988) suggested to eliminate them one per time, weighting the resulting columns in proportion to its number; e.g., if there are two redundant distributions, each one of the columns will weigh 0.5, and if having three, 0.33. Brooks (1990) and Brooks and McLennan (1991) proposed another strategy for dealing with parallelisms (dispersal events) that represent falsifications of the null hypothesis. It is named "secondary Brooks parsimony analysis" and consists of duplicating the involved area and dealing with each of the resulting areas separately. The analysis of the data matrix allows determining if it was really a unique area or if they were different areas incorrectly treated as a single one (Lomolino et al. 2006). Lieberman (1997, 2000,

2003, 2004) proposed another modification of BPA, named "modified BPA," intended to interpret geodispersal within a cladistic biogeographic framework. It involves two separate analyses: one to retrieve congruent episodes of vicariance and another to retrieve congruent episodes of dispersal, known as geodispersal (Lieberman 2004). The vicariance analysis produces a cladogram that makes predictions about the relative sequence of vicariance events that fragmented the areas. The geodispersal analysis produces a cladogram that provides information about the relative sequence of dispersal events that joined the areas. Both cladograms provide complementary information about the biotic assembly and can be placed within a geological framework. The procedure implies optimizing the ancestral states in the area cladograms, in order to estimate whether distributions implied expansions or contractions of the ancestral areas, and building the vicariance and geodispersal data matrices, following the same procedure as BPA. The best supported patterns of vicariance and geodispersal emerge from the parsimony analysis of these matrices. If both cladograms are relatively similar, the same geological processes may have produced vicariance and geodispersal (Lieberman 2004), e.g., cyclical sea-level rise and fall. If the cladograms are very different, they may imply that vicariance and geodispersal have been caused by not cyclical processes, e.g., continental collisions.

Brooks parsimony analysis has received some criticisms. Cracraft (1988) considered that BPA relies upon a questionable analogy to methods in systematics, so it has the potential to obscure the history of a biota rather than reveal it. For some authors, it tends to overestimate dispersal and extinction events (Dowling 2002). Enghoff (2000) criticized that BPA sometimes groups areas based on absent taxa. Ebach and Edgecombe (2001) noted that when taxa are mapped on the general area cladogram, anomalous reconstructions may appear, as descendants dispersing along with their ancestors, thus requiring a posteriori interpretations. Ebach et al. (2003) found that BPA sometimes gives spurious results. On the other hand, it has been argued that the parsimony principle should be used for analyzing the data and not for interpreting the results (Page 1989; Carpenter 1992). Miranda Esquivel et al. (2003) concluded that the events and duplication of areas in secondary BPA are ad hoc, so this method raises a scheme of verification and not of falsification. Siddall and Perkins (2003) compared the performance of BPA and tree reconciliation obtained with the software TreeMap, finding that sometimes BPA gives less parsimonious results. Siddall (2005) concluded that Brooks parsimony analysis lacks an optimality criterion and the coherence of a research program, because published descriptions of the methodology are self-contradictory. Furthermore, he considered that rules for a posteriori duplication of entities in secondary BPA are not specified clearly and that both primary and secondary BPA arrive at solutions that may defy a temporally consistent interpretation. Parenti (2007) criticized the codification strategy used to deal with geodispersal in modified BPA, because it specifies a direction that makes it an extension of Hennig's progression rule. Brooks et al. (2001, 2003) have responded to the criticisms considering that all the authors have applied the method incorrectly, because they did not take into account the modifications that were done to it after the original formulation.

These authors argued that primary BPA finds the most parsimonious general area cladogram, indicating in form of homoplasy how the null hypothesis of vicariance may be falsified. Secondary BPA integrates the incongruent elements, choosing the general area cladogram that postulates the smallest number of duplicated areas, each one of which represents a falsification of the null hypothesis.

The algorithm of primary BPA (Wiley 1987, 1988a, b; Dowling 2002) comprises the following steps (Fig. 2.11):

1. Obtaining the taxonomic cladograms of the taxa distributed in the areas analyzed.
2. Replacing the terminal species in the taxonomic cladograms by the areas inhabited by them, to obtain taxon-area cladograms.
3. Labeling the components as well as the widespread terminal species (assumption 0) in the taxon-area cladograms.
4. Constructing a data matrix where areas are the rows, and components and widespread terminal species the columns, coding "1" if the area is present and "0" if it is absent. Use "?" for missing areas. Add a row with all "0" to root the cladogram.
5. Analyzing the data matrix with a parsimony algorithm, in order to obtain the general area cladogram.
6. Optimizing the components in the general area cladogram, to identify vicariance events (= synapomorphies), dispersal events (= parallelisms), and extinctions (= reversals).

A completely different method is parsimony analysis of paralogy-free subtree analysis. It is based on the concept of paralogous areas, which are those areas that conflict with duplications of themselves. Nelson and Ladiges (1996, 2001) considered that geographic paralogy causes that the components that may provide biogeographic information are not directly informative. This means that we may have contradictory relationships, due to sympatric speciation, lack of response to vicariance events, and incorrect definition of areas and other explanations, which can lead to erroneous interpretations (Nelson and Ladiges 2001). Paralogy-free subtrees simplify the cladistic biogeographic analysis, so that geographic data need not be associated with paralogous nodes, preventing artefactual results, if not altogether at least to a significant degree (Nelson and Ladiges 2003; Parenti 2007).

Nelson and Ladiges (1996) developed an algorithm that constructs paralogy-free subtrees, starting off at the most terminal groups of the cladogram. The procedure reduces complex cladograms to paralogy-free subtrees, meaning that geographic data are associated only with informative nodes, and areas duplicated or redundant in the descendants of each node do not exist. These are the only data relevant for cladistic biogeography. Once obtained, paralogy-free subtrees are represented in a component or a three-item matrix and analyzed with a parsimony algorithm. Prior to obtaining of paralogy-free subtrees, the transparent method (Ebach et al. 2005) may be implemented to resolve widespread taxa.

The algorithm of parsimony analysis of paralogy-free subtrees (Morrone 2014c) comprises the following steps (Fig. 2.12):

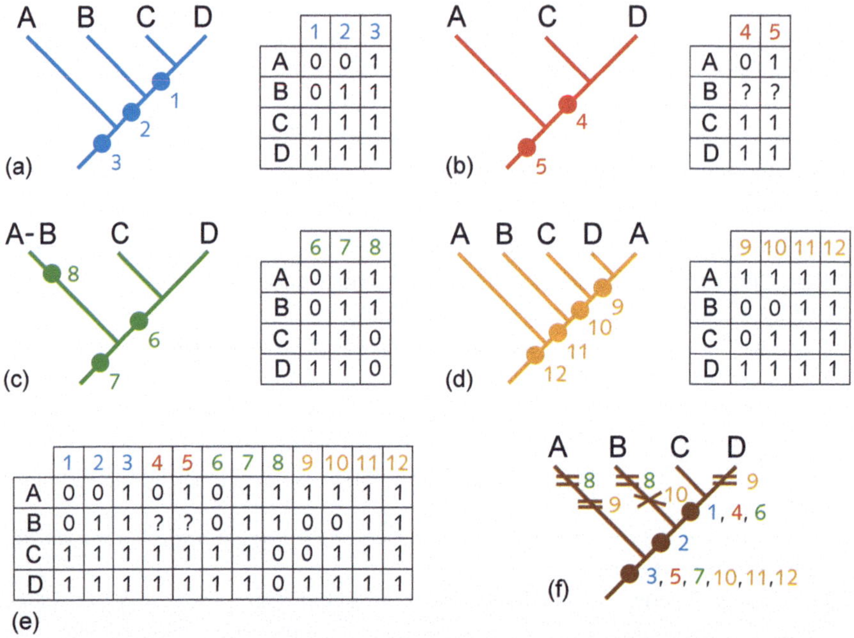

Fig. 2.11 Brooks parsimony analysis (BPA). (**a**–**d**) taxon-area cladograms and partial matrices derived from them; (**a**) trivial case; (**b**) taxon with a missing area; (**c**) widespread taxon; (**d**) taxon with a redundant distribution; (**e**) data matrix with all the information; (**f**) general area cladogram obtained

1. Obtaining the taxonomic cladograms of the taxa distributed in the areas analyzed.
2. Replacing the terminal taxa from the taxonomic cladograms by the areas inhabited by them, to obtain taxon-area cladograms.
3. Resolving widespread taxa with the transparent method and identifying the paralogy-free subtrees starting at each terminal node and progressing to the base of each taxon-area cladogram. When a node leads to one or more terminal taxa that are geographically widespread, and part of that distribution overlaps with that of another taxon or taxa, reduce the widespread distribution to the nonoverlapping geographic element.
4. Representing the nodes of all the paralogy-free subtrees in a data matrix.
5. Analyzing the data matrix with a parsimony algorithm to obtain the general area cladogram.

2.4 Biogeographic Regionalization

One of the most striking facts of the geographic distributions of taxa is that they have limits, and since these limits are repeated for different taxa, they allow the recognition of biotas (Morrone 2009, 2018). Small biotas are nested within larger

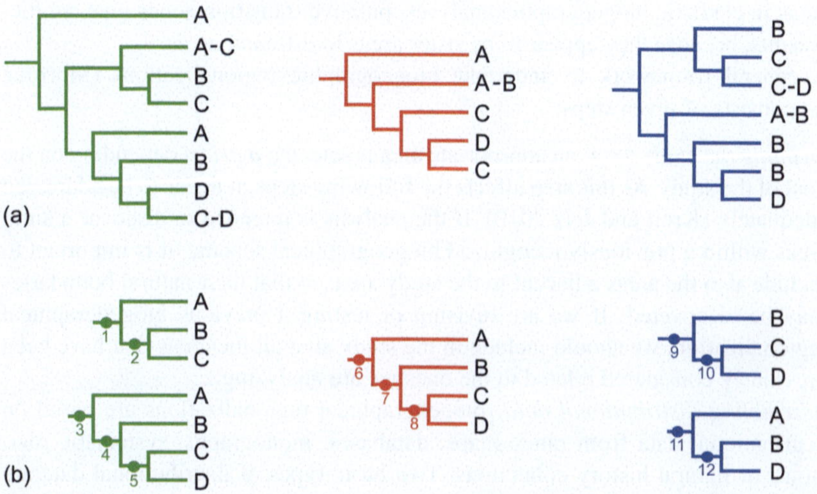

	1	2	3	4	5	6	7	8	9	10	11	12
A	1	0	1	0	0	1	0	0	0	0	1	0
B	1	1	1	1	0	1	1	0	1	0	1	1
C	1	1	1	1	1	1	1	1	1	1	0	0
D	0	0	1	1	1	1	1	1	1	1	1	1

(c)

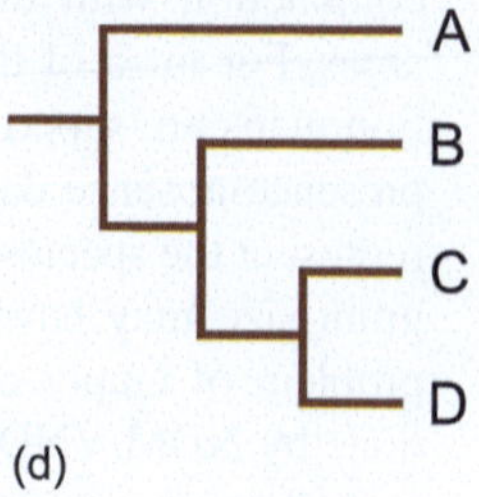

Fig. 2.12 Parsimony analysis of paralogy-free subtrees. (**a**) Original taxon-area cladograms, with paralogous nodes; (**b**) paralogy-free subtrees that are derived from them; (**c**) data matrix; (**d**) general area cladogram obtained

biotas, so they can be ordered hierarchically in a system of kingdoms, regions, dominions, provinces, and districts. Given the historical and logical primacy of classification over process explanations (Rieppel 1991, 2004), this stage of the analysis takes place before cenocrons are elucidated and a geobiotic scenario is proposed.

Sometimes it is difficult to determine the exact boundaries of two kingdoms or regions. For example, Müller (1979) illustrated 18 proposals of boundaries between the Palearctic and Ethiopian regions. As a result, authors have described transition zones (Darlington 1957; Halffter 1987), which represent events of biotic "hybridization," promoted by historical and ecological changes that allowed the mixture of different biotas. Transition zones may have a depauperate biota, but in some cases, they harbor a particularly high biodiversity (Ferro and Morrone 2014). In track analyses, transition zones are detected by the presence of nodes (Escalante et al. 2004), whereas in cladistic biogeographic analyses, putative transition zones give conflictive results, because they appear to be sister areas to different areas.

A general framework to undertake biogeographic regionalizations (Morrone 2018) consists of seven steps:

1. *Defining the study area:* an appropriate area is selected *a priori* depending on the goal of the study. As this step affects the following steps, it needs to be addressed adequately (Kreft and Jetz 2010). If the analysis is intended to discover natural areas within a previously recognized biogeographical scheme, it is important to include also the areas adjacent to the study area, so that their natural boundaries may be discovered. If we are revising or testing a previous biogeographical regionalization, we should include in the study area all the areas that have been previously considered related to the ones we are analyzing.
2. *Assembling distributional data:* biogeographical regionalizations are based on distributional data from range maps, databases, monographs, systematic revisions, or natural history collections. Two basic types of distributional data are used for biogeographical analyses: extent-of-occurrence maps and point information. Extent-of-occurrence maps are based on the opinion of experts and are represented on maps as polygons. Point information may be used by itself or in conjunction with distributional modeling techniques to depict distributional areas. For most of the analyses, extent-of-occurrence maps and point information maps are superimposed to a grid. Then, grid cells can be transferred into a presence-absence data matrix, where rows represent the grid cells and columns represent the species. The use of grid cells is not without problems, for example, grain size may have great relevance (see Morrone and Escalante 2002). The problem of empty cells can be overcome by a "moving window" strategy, as done by NDM/VNDM (Goloboff 2011). In addition to grid cells, there are several alternative area units that have been explored (Morrone 2009). Which taxa should be analyzed? Although many of the first authors providing world regionalizations used supraspecific taxa when recognizing large areas as kingdoms or regions, it has been common in most recent studies to analyze only species. The problem with this approach is that species usually allow to discover relatively small areas (e.g., those treated as districts or provinces), so some authors have incorporated information on supraspecific taxa, either using phylogenetic information (e.g., Holt et al. 2013) or by considering explicitly families and genera (e.g., Escalante 2017).
3. *Identifying natural areas:* an area of endemism is identified by the co-occurrence of two or more endemic taxa. It is usually limited by geographical barriers, altitudinal ranges, or a vegetation type. This distributional pattern has been considered to represent a statement of primary biogeographic homology (Morrone 2001), which refers to a conjecture on a common biotic history, based on the co-distributional patterns of different plant and animal taxa. Areas based on ecological grounds, known as ecoregions, may be recognized, and in occasions, they can be equivalent to areas of endemism based on endemic taxa. Biogeographical areas are commonly mapped as contiguous, in opposition to ecoregions, which

are usually represented as discontinuous. To identify meaningful areas, and especially when dealing with large data sets, a quantitative method should be selected. There are different alternative procedures: methods using similarity indices and clustering techniques (Kreft and Jetz 2010), parsimony analysis of endemicity (Morrone 2014b), endemicity analysis (Szumik et al. 2002), geographical interpolation (Oliveira et al. 2015), and associational networks (Vilhena and Antonelli 2015). There are a few analyses comparing some of these methods, but there is no consensus on which could be the most appropriate.

4. *Discovering area relationships:* biogeographical regionalizations have a hierarchical structure (Escalante 2009), where smaller areas are nested within larger ones. To provide such a hierarchy, different strategies have been proposed. Olivero et al. (1998) explored the use of strong and weak boundaries for providing such hierarchy. Holt et al. (2013) considered incorporating phylogenetic relationships to quantify the affinities among regions more appropriately and, at the same time, evaluating the spatial turnover in the phylogenetic composition of the biotas analyzed. The area relationships implicit in this hierarchy represent a shared biotic history, and cladistic biogeography is the approach specifically designed to discover biotic relationships based on the phylogenetic relationships of the taxa analyzed (Parenti and Ebach 2009). Phylogenetic analyses for different taxa are currently being published at an extraordinary rate, so the possibilities of undertaking cladistic biogeographical analyses are enormous, although there is no agreement on which is the most appropriate method (Morrone 2009).
5. *Defining the boundaries and transition zones:* defining boundaries between different biogeographical regions is not straightforward. Usually, different taxa show different boundaries, so a unique line cannot be drawn, but instead a transition zone is represented. Transition zones are geographical areas of overlap, with a gradient of replacement and partial segregation between different biotas (Ferro and Morrone 2014). Olivero et al. (2011) developed a procedure based on fuzzy sets that in addition to identifying natural areas may be used to discover transition zones, assuming that boundaries between regions are usually not sharply defined but consist of broad transition zones. Transition zones have been identified in the areas of biotic overlap of different biogeographical kingdoms or regions; their recognition at lower hierarchical levels has been suggested occasionally (e.g., Morrone 2004b, 2006; Escalante 2009). There are no objective criteria to consider any biogeographic unit as belonging to a region or a transition zone. Roig-Juñent et al. (2018) proposed recently that a biogeographic province should be considered part of a region when one of its biotic elements surpasses 70%, whereas biogeographic provinces without predominance of any biotic element should be treated as transitional.
6. *Regionalization:* to reflect the hierarchical organization of the areas recognized, a system of categories is applied. The most commonly used categories are kingdom, region, dominion, province, and district (Ebach et al. 2008; Ebach and Parenti 2015). If necessary, intermediate categories with the prefix "sub" may be used, e.g., subregions, subprovinces, etc.

7. *Area nomenclature:* some biogeographical regionalizations have followed the nomenclatural conventions set out in the International Code of Area Nomenclature or ICAN (Ebach et al. 2008). ICAN provides a universal naming system to standardize area names used in biogeography and other disciplines, where names are grouped under more inclusive area names to represent a biogeographical hierarchy (kingdoms, regions, dominions, provinces, and districts). The notion of priority is applied to use the oldest available names instead of new names. The work of Sclater (1858) is adopted as the date of the starting point of biogeographical nomenclature, as it constitutes the first widely adopted world biogeographical regionalization; in some cases, widely used names were kept instead of older synonyms, applying a criterion analogous to the *nomen conservandum* convention of taxonomical nomenclature, to provide a better stability (Morrone 2017).

2.5 Identification of Cenocrons

Several authors have postulated that dispersal explanations reside on narrative frameworks, lacking a general theory to explain distributional patterns, representing ad hoc explanations (e.g., Croizat 1958; Croizat et al. 1974; Nelson and Platnick 1981). After establishing secondary biogeographic homology patterns in cladistic biogeographic analyses (Morrone 2001, 2009), dispersal explanations can help establish when the cenocrons assembled in the identified biotas, incorporating a time perspective to the study of biotic evolution. This time perspective may be incorporated by time-slicing, intraspecific phylogeography, and molecular dating. Based on the information provided by time-slicing, intraspecific phylogeography, and molecular dating, it is possible to provide a time framework for the biotic subsets that may be considered cenocrons (Morrone 2009). This time perspective, in addition to the current distribution of each taxon, the current geographical distribution of its sister taxon (or related taxa, in case of unresolved phylogenies) and the phylogenetic relationships of the higher taxon where it belongs, allows to hypothesize on the existence of a cenocron (Corral-Rosas and Morrone 2017; Roig-Juñent et al. 2018).

Cenocrons are evolutionary units, but there are some ecological aspects that deserve to be considered (Lobo 1999, 2007; Halffter and Morrone 2017). Some adaptations of the taxa belonging to a cenocron may have been acquired by their ancestors in the areas where they originally evolved, and some of these adaptations may be relevant when explaining their current distribution. Lobo (1999) has named these adaptations "biogeographic memory" and Halffter and Morrone (2017) "ecological inertia." The species belonging to the same cenocron have suffered a similar history of dispersal and assembly into a biota, although they may have not had the same temporal origin. Thus, when analyzing the species that hypothetically belong to a particular cenocron, we should identify those species that, event belonging to different lineages, have been able to colonize an area possessing the extreme conditions of the gradient where these lineages can survive (Lobo 2007). Based on this ecological framework, Lobo (2007) considered that the identification of cenocrons should also consider information on the current abiotic conditions of the areas where they are distributed.

Time-Slicing Hunn and Upchurch (2001) have emphasized the relevance of time in evolutionary biogeography, because data on the temporal distribution may provide important constraints in biogeographic analyses, helping reinforce or overturn specific hypotheses. Lieberman (2004) has highlighted the relevance of the deep time perspective in biogeography, postulating that the fossil record is the only primary and direct chronicle of the history of life, the phenomenon of extinction can influence our ability to retrieve biogeographic patterns, and the fossil record may provide examples of dispersal events. It has been suggested that the temporal ranges of organisms are important because distribution patterns seem to "decay" through time as new ones are superimposed (Grande 1985; Upchurch and Hunn 2002; Upchurch et al. 2002). Donoghue and Moore (2003) postulated that cladistic biogeographic methods are susceptible to the confounding effects of pseudo-incongruence and pseudo-congruence, if they do not incorporate information on the absolute timing of the diversification of the lineages. Pseudo-incongruence occurs when different area cladograms show conflict because taxa evolved at the same time but diversified in response to different events. Pseudo-congruence occurs when different area cladograms show the same area relationships, although the taxa diversified at different times, presumably under different underlying causes.

Cladistic biogeographers have usually avoided using temporal data because of the risk of incorporating ideas of unobserved processes in the elucidation of biogeographic patterns (for an alternative view, see Folinsbee and Evans 2012). This would imply unverifiable assumptions, with the risk of falling back on narrative scenarios. The need for considering time in biogeography, however, becomes clearer in cases of biogeographic convergence. The terms convergence and divergence have been proposed by Hallam (1974) to distinguish two extreme biogeographic patterns. Widespread taxa and redundancy identify biogeographic convergence, whereas vicariance is the most common interpretation of divergence patterns. Convergence can be the result of dispersal or area coalescence due to the elimination of geographic barriers. Analyses of biogeographic convergence are unlikely to show congruence. Upchurch et al. (2002) noted that biogeographic analyses over an extensive stratigraphical range may fail to find the correct area relationships.

This point is illustrated by the hypothetical succession of area separations followed by area coalescence described by Upchurch and Hunn (2002), where branching relationships only are evident when extensive dispersal has not yet overwhelmed the original vicariance pattern. Area coalescence causes vicariance patterns to fade out through time (Grande 1985), and analytical techniques based on parsimony algorithms may be inappropriate (Young 1995). When previously merged biotas subsequently undergo vicariance, patterns may need to be treated within a multiple time-plane approach (Rosen and Smith 1988) or time-slicing (Upchurch et al. 2002). The latter term corresponds to the analysis of biotic distributional data according to a sequence of individual stratigraphical intervals or time-slices.

The solution to problems posed by instances of biogeographic convergence is time-slicing (Grande 1985; Upchurch and Hunn 2002; Cecca et al. 2011). While assessments of faunal similarity are usually undertaken with faunas of successive geological ages, traditional cladistic biogeography has only used data on organism relationships and spatial distributions on a single time plane (usually the present). Time-slicing may reconcile the use of time and a synchronic approach. Ideally, paleobiogeographers should be able to use a synchronic approach for each time-slice they identify. This is difficult because of the limits imposed by geological constraints, e.g., insufficient precision or resolution of chronological correlations, incompleteness of the fossil record, etc. Upchurch and Hunn (2002) and Upchurch et al. (2002) proposed temporally partitioned component analysis (TPCA), also known as chronobiogeography, to incorporate explicitly temporal data into a cladistic biogeographic analysis (for an alternative approach see Folinsbee and Evans 2012). After geodispersal, due to area coalescence events (biogeographic convergence), "the histories of areas and biotas will have a reticulated rather than a branching structure, raising the question as to how well cladistic biogeographic techniques will be able to accurately analyze and depict a reticulate system" (Upchurch and Hunn 2002, p. 280). The starting point of TPCA is the existence of taxon-area cladograms for the taxonomic groups on which the analysis is based. Ideally, synchronic relations would be found for each time-slice on the basis of phylogenetic relations, allowing reticulate histories to be explained, where assumptions are relatively minimized.

Temporally partitioned component analysis (Upchurch and Hunn 2002) comprises the following steps:

1. Pruning or temporally partition the taxon cladograms by deleting all taxa that did not exist at a particular designated time-slice
2. Finding optimal area cladograms for each particular time-slice, by determining which area relationships provide the best (under some designated optimality criterion) explanation for the spatial distributions observed in the taxon cladogram
3. Using a randomization test to determine whether the degree of fit between area and taxon cladogram for each time-slice is greater than would be expected by chance.

In recent years, some alternatives for time-slicing have been formulated (Corral-Rosas and Morrone 2017; Gámez et al. 2017; King and Ebach 2017). King and Ebach (2017) considered that time-slicing areas rather than the phylogenetic trees may identify temporally overlapping areas. They analyzed taxa distributed in the Indo-Malayan Transition Zone, dating them into Paleogene (>23 m.y.a.) and Neogene (<23 m.y.a.). Areas of endemism were selected based on paleogeographic reconstructions for the Paleogene and based on current distributional data for the Neogene. The analysis of the sliced areas showed that Wallace's line (see Fig. 1.1) represents a Paleogene barrier, Weber's or Mayr's line represents a Neogene barrier, and Lydekker's line does not represent a boundary. Thus, the Indo-Malayan Transition Zone is a temporally composite area.

Intraspecific Phylogeography Phylogeography studies the geographic distribution of genealogical lineages, especially those within and among closely related species, based on molecular data (Avise et al. 1987; Avise 2000; Lanteri and Confalonieri 2003; Lomolino et al. 2006). The maternal, nonrecombinant mode of inheritance of mitochondrial DNA (mtDNA) and the rapid evolution of mtDNA sequences make it possible to obtain haplotypes (combinations of alleles at multiple linked loci), which can be used to obtain intraspecific phylogenetic hypotheses (Crisci et al. 2003). Once the population genetic structure has been assessed based on mtDNA, it is possible to obtain a network or cladogram of haplotypes, which allows to analyze historical patterns and the processes that have shaped them, e.g., dispersal, vicariance, range expansion, and colonization, sometimes under a statistical framework (Knowles and Maddison 2002; Templeton 2004). Phylogeographic analyses are basically constrained to Pleistocene and Holocene time frames, with some studies extending back into the Neogene (Riddle et al. 2008). Paleophylogeography (Betancourt 2004) has allowed to find phylogeographic structure in fossil populations; the analysis of Kuch et al. (2002) of ancient cytochrome b fragments from rodent fecal pellets from the Atacama Desert (Chile) provided interesting insights on paleophylogeography.

The results of a phylogeographic analysis are usually represented as a parsimony network connecting the studied haplotypes, where the number of restriction sites differences indicates the relative distance between haplotypes and groups of haplotypes. When mapped, disjunctions and sympatry between the haplotypes may give us clues on their evolutionary histories. Highly divergent groups inhabiting disjunct areas can indicate independent evolutionary histories for a relatively long period of time, usually due to vicariance, whereas lack of geographic structure may indicate recent dispersal. In animal species, mtDNA has proven to be useful for phylogeographic studies, but in plants, it evolves quickly with respect to gene order but slowly in nucleotide sequence, so it is of limited utility (Crisci et al. 2003). Chloroplast DNA (cpDNA) has shown to be structured geographically in several plant species; it is also transmitted maternally and has exhibited considerable intraspecific variation (Soltis et al. 1992; Avise 2000, 2004).

An explicit genealogical approach to intraspecific phylogeography implies extending the lineage sorting theory to sister populations (Avise 2000). In terms of maternal genealogy, there are three possible situations: polyphyly, where some but not all the extant matrilines in one population join with some but not all the extant matrilines in the other to form a clade; paraphyly, where all the matrilines within one population from a clade nested within the broader matrilineal history of the other population; and reciprocal monophyly, when all the extant matrilines within each sister population are closer genealogically to each other than to any matriline in the other. These three situations may characterize the same pair of sister populations at different time depths. Three basic phylogeographic hypotheses have been formulated by Avise et al. (1987):

1. Most species are composed of geographic populations whose members occupy recognizable matrilineal branches of an extended intraspecific pedigree, displaying significant phylogeographic structure supported by mtDNA data.
2. Species with limited or "shallow" phylogeographic population structure have occupied ranges free of long-standing barriers to gene flow, exhibiting limited phylogeographic structure.
3. Intraspecific monophyletic groups distinguished by large genealogical gaps usually arise from long-term extrinsic biogeographic barriers to gene flow. Major phylogeographic units within a species reflect long-term historical barriers to gene flow. This hypothesis has four corollaries representing four aspects of genealogical agreement: agreement across sequence characters within a gene, agreement in significant genealogical partitions across multiple genes within a species agreement in the geography of gene-tree partitions across multiple codistributed species, and agreement of gene-tree partitions with spatial boundaries between traditionally recognized biogeographic units.

Five different phylogeographic patterns (Avise 2000) can be characterized for mtDNA gene cladograms:

Deep gene tree, lineages allopatrid: it is characterized by the presence of spatially circumscribed haplotypes, separated by relatively large mutational distances. This pattern appears commonly in phylogeographic patterns of mtDNA. A long-term extrinsic barrier to genetic exchange is the most commonly invoked explanation.

Deep gene tree, lineages broadly sympatrid: it is characterized by pronounced phylogenetic gaps between some branches in a gene tree, with main lineages codistributed over a wide area. It could arise in a species where some anciently separated lineages might have been retained by chance, whereas many intermediate lineages were lost over time by gradual lineage sorting.

Shallow gene tree, lineages allopatrid: most or all haplotypes are related closely, yet are localized geographically. Contemporary gene flow has been low enough in relation to population size to have permitted lineage sorting and random drift to promote genetic divergence among populations that were in historical contact recently.

Shallow gene tree, lineages sympatrid: it is expected for high-gene flow species of small effective size whose populations have not been sundered by long-term barriers.

Shallow gene tree, distributions varied: intermediate between the two previous categories, it involves common lineages that are widespread plus closely related lineages that are confined to one or a few nearby localities. It implies low contemporary gene flow between populations that are connected tightly in history. Common haplotypes are often plesiomorphic, and rare haplotypes are presumably apomorphic.

Phylogeographic patterns exhibiting large genetic gaps between phylogroups (groups of closely related haplotypes separated from other phylogroups by relatively large genetic gaps), such as those represented in the two first categories are amenable to phylogenetic analyses (Lomolino et al. 2006). This can be done by simply treating each phylogroup as a terminal unit in the cladistic analysis. When variable haplotypes are closely related and groups of populations are not clustered within the clearly reciprocally monophyletic phylogroups, such as those in the three latter categories, one can still present an unrooted cladogram to summarize haplotype relationships.

Molecular Dating The assumption that the rate of molecular evolution is constant over time for proteins allows inferring a clock-like accumulation of molecular changes (Zuckerland and Pauling 1965; Bromham and Woolfit 2004; Morrone 2009). The "ticks" of the molecular clock (corresponding to mutations) do not occur at regular intervals but rather at random points in time (Gillespie 1991). This time is measured in arbitrary units and then calibrated in millions of years by reference to the fossil record or geological data (Sanderson 1998; Magallón 2004; Benton and Donoghue 2007), giving minimum estimates of the age of a clade, which in turn may help elucidate the relative minimum ages of the cenocron to which it belongs. Additionally, estimates of relative minimum ages of divergence may help draw conclusions about historical processes that may have affected the taxa and decide whether a specific dispersal or vicariance event hypothesis better explains the patterns observed (Riddle et al. 2008). If calibrations provide estimates smaller than those proposed by vicariance events, dispersal may be a better explanation (Morrone 2009).

The calibration of a molecular clock requires that we find two extant species for which the date of speciation can be determined from the fossil record, to establish the time since the speciation event. Then, we compare the DNA sequences of the same gene of both species and count the number of nucleotide substitutions. If all the substitutions are assumed to have arisen subsequently to the speciation event, the rate of DNA evolution for the gene under study is obtained by dividing the number of DNA differences between both species by the time since speciation. Assuming a constant mutation rate, we can extrapolate the approximate dates of speciation for other species, for which no fossil dates are available (Crisci et al. 2003). In order to test the molecular clock hypotheses, there are three available tests: the likelihood ratio test, the dispersion index, and the relative rate test (Page and Holmes 1998).

As analyses from several taxa began to accumulate, it became apparent that the molecular clock is not always a good model for the process of molecular evolution (Rutschmann 2006). If the null hypothesis of a constant rate is rejected or if we have evidence suggesting that rates vary across the tree, we may have to use methods that correct for rate heterogeneity (Rambaut and Bromham 1998) or that estimate divergence times by incorporating rate heterogeneity (Kishino et al. 2001; Sanderson 2002). Other problems associated with molecular methods include the stochastic nature of molecular substitution, the assumption of rate constancy among lineages when such constancy is absent, and the link between substitution rate and elapsed time on the branches of a cladogram (Magallón 2004). Calibration made by

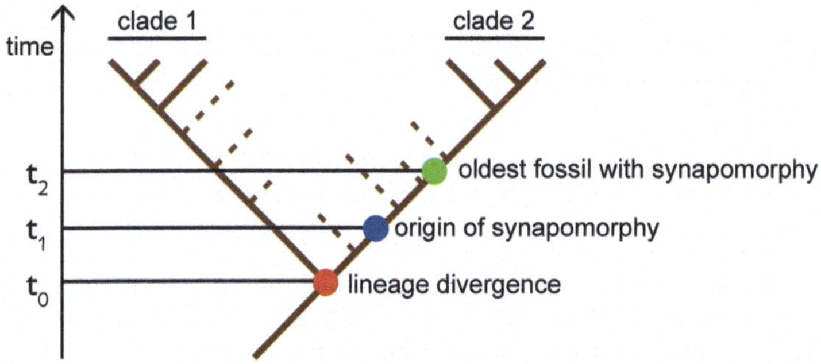

Fig. 2.13 Temporal gap between the divergence of a taxon and its sister taxon (t_0), the origin of the synapomorphy (t_1) and the occurrence of the oldest fossil bearing such synapomorphy (t_2)

reference to geological events runs the risk of circular reasoning, because the clock is used to test biogeographic hypotheses which involve an event potentially caused by a geological process (Crisci et al. 2003). Bromham and Woolfit (2004) noted that sometimes molecular date estimates are notoriously at odds with other lines of evidence, e.g., the dates of the "Cambrian explosion" of Metazoan phyla, and radiations of mammals and birds are almost twice as old as the available fossils suggest. This discrepancy may be due to systematic biases in the fossil record that left particular taxa, regions or periods effectively unrecorded, or because explosive radiations could speed the molecular clock, causing dates for these radiations to be consistently overestimated. Magallón (2004) and Benton and Donoghue (2007) clarified the relationship of the fossil record and molecular dating methods, the former documenting first appearances of morphological features, and the latter dating splits of molecular lineages (Fig. 2.13).

There are several molecular dating methods, grouped into three main classes: methods that use a molecular clock and one global rate of substitution, those that correct for rate heterogeneity, and those that try to incorporate rate heterogeneity. Each method has its own algorithm. For a revision, see Rutschmann (2006).

Heads (2004, 2005, 2009, 2011) provided a critique of molecular dating, proposing instead that lineages should be calibrated using tectonic events separating sister-taxa and that fossils should be used to provide minimum, not maximum, ages. Swenson et al. (2012) criticized the idea of calibrating molecular phylogenies *a priori* using sister-group relationships as representing true vicariance events. They found that extrapolating dates based on tectonic calibrations could lead to incorrect results.

Horobiotas Once a cenocron is incorporated to a biota, the resulting assemblage may be considered a horobiota (Reig 1981). A horobiota represents a set of species belonging to different cenocrons that coexist and diversify during an extended lapse, thus representing a lasting biogeographic unit.

2.6 Construction of a Geobiotic Scenario

Once we have identified the biotas and their cenocrons, we may be able to construct a geobiotic scenario by accounting biological data (means of dispersal, etc.) as well as nonbiological data (past continental configurations, etc.). These data allow to integrate a plausible scenario to explain the episodes of vicariance and dispersal that have shaped biotic assembly.

Evolutionary biogeographers have shown great interest in geology, geophysics, and plate tectonics (Craw 1988; Cooper 1989; Heads 1989; Michaux 1989; Craw et al. 1999; Grehan 2001a). Geology and biogeography have a causal relationship, being the independent and dependent variables, respectively (Michaux 1989). This does not imply that geological hypotheses necessarily validate biogeographic hypotheses, because geologists may have not necessarily interpreted geological history of the area adequately enough to justify validation. In fact, the relationship between geology and biogeography should be based on its capacity of "reciprocal illumination" (in the sense of Hennig 1950 and Santos and Capellari 2009). In order to make this relation more fruitful, it would be important to develop a common language to interconnect the biological and geological systems. This is because evolution in space/time of biotas is a unique geobiotic phenomenon. Generalized tracks and general area cladograms constitute appropriate instruments to develop such common language (Morrone 2009; Echeverry et al. 2012).

Biogeographers have classified geographic features in terms of their impact on dispersal and vicariance (Simpson 1953; Rapoport 1975; Vargas 1992; Cox and Moore 1998; MacDonald 2003). The most important are barriers (geographic features that hinder dispersal) and corridors (geographic features that facilitate dispersal). Barriers are easily identified with geographic elements as mountains, rivers, seas, etc. In the marine environment, in addition to land barriers (e.g., the Isthmus of Panama), there are more subtle barriers, represented by changes in physicochemical properties (Cecca 2002). Corridors include a variety of habitats that many organisms found at either end of them find little difficulty in traversing them (Cox and Moore 1998). These terms are relative, because, for example, a cordillera may act as a barrier for certain species but be a corridor for others. Instead of barriers, cladistic biogeographers usually refer to vicariance events. In some instances, physical or biological conditions make it easier or more difficult for certain species to cross a certain barrier. Features that are not equally favorable for dispersal of all species are called filters. For example, before the rise of the Isthmus of Panama, there was a chain of islands (stepping stones) set upon a relatively shallow sea of about 150 m in depth occupying its place. These islands acted facilitating dispersal of some species but acted as a barrier for other species. After the Isthmus of Panama developed during the Pleistocene, most of Central America was occupied by dense tropical forests, which allowed the dispersal of forest species but acted as a barrier for biota from the savannas (MacDonald 2003). There are some areas completely surrounded by totally different environments, like islands, caves, or high mountain peaks, where chances of dispersal are very low for most of the taxa. They are known as sweepstake routes, and differ from filters in kind, not merely in degree, for almost all the species that traverse them cannot survive (Cox and Moore 1998).

When dealing with long-term changes in the biotic distributional patterns, continental drift may be a relevant factor (Briggs 1987; Cox and Moore 1998). The splitting and collision of land masses not only affect distributional patterns directly but also new mountains, oceans, or land barriers change climatic patterns upon the land masses. Continental drift was originally proposed by Wegener (1912) and found enormous opposition. Plate tectonics was the mechanism that explained continental drift and made it a credible theory. Seafloor spreading is caused by great convection currents that bring material to the surface from the hot interior of the Earth, inducing the movement of the tectonic plates. These constitute the moving units at the surface of the Earth and may contain continental masses or may consist of ocean floor. The movement of the plates had great relevance for the organisms on Earth. The movement of the continents relative to the poles and the equator caused climatic changes. Additionally, shallow epicontinental seas covered parts of the continents or formed seas within them during the Jurassic and Cretaceous periods, forming barriers to dispersal. The splitting of continents also altered the patterns of water circulation in the oceans.

Sanmartín and Ronquist (2004) provided a synthesis of the ideas concerning the fragmentation of Gondwana, which they presented as a geological area cladogram (Fig. 2.14). Gondwana started to breakup in the Jurassic (165–150 m. y. a.), when rifting between India and Australia-east Antarctica began. Then, Madagascar-India broke away from Africa and began moving southeast, attaining its present position in the Early Cretaceous (121 m. y. a.). India separated from Madagascar in the Late Cretaceous (88–84 m. y. a.) and began drifting northward, colliding with Asia *ca.* 50 m. y. a. South America began to separate from Africa in the Early Cretaceous (135 m. y. a.). Northern South America and Africa remained connected until the Mid-Late Cretaceous (110–95 m. y. a.), when a transform fault opened between them, and Africa started drifting northeast and collided with Eurasia in the Paleocene (60 m. y. a.). New Zealand, Australia, South America, and Antarctica remained connected until the Late Cretaceous. About 80 m. y. a., the Tasmantis block (New Zealand plus New Caledonia) broke away from west Antarctica and moved northwest, opening the Tasman Sea. New Zealand and New Caledonia finally separated in the Neogene (40–30 m. y. a.). Australia and South America remained in contact across Antarctica until the Eocene. Australia began to separate from Antarctica in the Late Cretaceous (90 m. y. a.), but both remained in contact along Tasmania, and complete separation did not occur until the Late Eocene (35 m. y. a.). Southern South America and Antarctica remained in contact through the Antarctic peninsula until the Oligocene (20–28 m. y. a.). New Guinea then joined to the Australian plate, although only the southern margin of New Guinea was emergent at that time. The collision between the Australian and Pacific plates in the Oligocene (30 m. y. a.) initiated the tectonic uplift of New Guinea. The link between North and South America, the Isthmus of Panama, was formed in the Late Pliocene (2 m. y. a.).

There is still considerable discussion about some aspects of plate tectonics. Theories postulating a lost Pacifica continent (Kamp 1980; Nur and Ben-Avraham

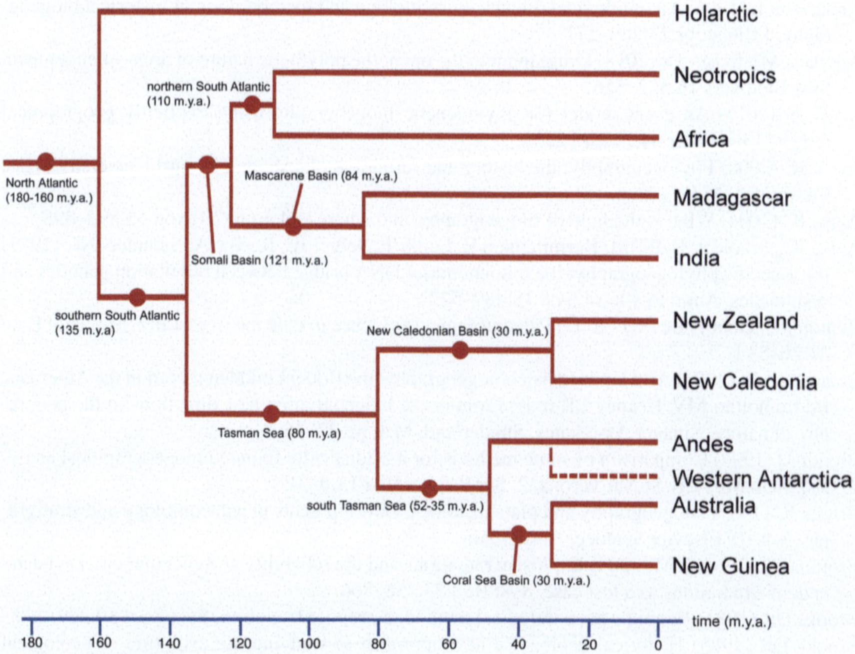

Fig. 2.14 Geological area cladogram presented by Sanmartín and Ronquist (2004), which represents the relationships among the southern continents based on paleogeographic evidence

1980) or an expanding Earth (Shields 1979, 1991, 1996; McCarthy 2003, 2007) have been proposed to explain certain "anomalies" of Wegener's theory, but they have not gained support by geophysicists (Humphries and Ebach 2004). Any of these theories implies a major role for vicariance in isolating populations of plant and animal species (Cox and Moore 1998). Continents split, and their fragments carry away its cargo of living organisms, known as "Noah's arks," and buried fossils, known as "Viking funeral ships" (McKenna 1973).

Other tectonic studies are more local. Echeverry et al. (2012) obtained a geological area cladogram for the Caribbean. It shows that a group of allochthonous tectonostratigraphic terranes with a Pacific origin assembled in the Caribbean plate. Halffter and Morrone (2017) provided a series of paleogeographical maps of North America showing in a temporal sequence the opportunities of dispersal and vicariance in the Mexican Transition Zone (see Chap. 5).

References

Aguilar-Aguilar R, Contreras-Medina R, Salgado Maldonado G (2003) Parsimony analysis of endemicity (PAE) of Mexican hydrological basins based on helminth parasites of freshwater fishes. J Biogeogr 30:1861–1872

Andersson L (1996) An ontological dilema: epistemology and methodology of historical biogeography. J Biogeogr 23:269–277

Apodaca MJ, Crisci JV (2018) Dragging into the open: the polythetic nature of areas of endemism. Syst Biodivers 16:522–526

Arias JS (2017) An event model for phylogenetic biogeography using explicitly geographical ranges. J Biogeogr 44:2225–2235

Avise JC (2000) Phylogeography: the history and formation of species. Harvard University Press, Cambridge, MA

Avise JC (2004) What is the field of biogeography, and where is it going? Taxon 53:893–898

Avise JC, Arnold J, Ball RM, Bermingham E, Lamb T, Neigel JE, Reeb CA, Saunders NC (1987) Intraspecific phylogeography: the mitochondrial DNA bridge between population genetics and systematics. Annu Rev Ecol Syst 18:489–522

Benton MJ, Donoghue PCJ (2007) Paleontological evidence to date the tree of life. Mol Biol Evol 24:26–53

Betancourt JL (2004) Arid lands in paleobiogeography: the Rodent midden record in the Americas. In: Lomolino MV, Heaney LR (eds) Frontiers of biogeography: new directions in the geography of nature. Sinauer Associates, Sunderland, MA, pp 27–46

Biondi M (1998) Comparison of some methods for a cladistically founded biogeographical analysis. Mem Mus Civ St Nat Verona, 2. Ser Sez Sci Vita 13:9–31

Briggs JC (1987) Biogeography and plate tectonics. Developments in palaeontology and stratigraphy, vol 10. Elsevier Science, Amsterdam

Bromham L, Woolfit M (2004) Explosive radiations and the reliability of molecular clocks: Island endemic radiations as a test case. Syst Biol 53:758–766

Brooks DR (1981) Hennig's parasitological method: a proposed solution. Syst Zool 30:229–249

Brooks DR (1985) Historical ecology: a new approach to studying the evolution of ecological associations. Ann Missouri Bot Gard 72:60–680

Brooks DR (1990) Parsimony analysis in historical biogeography and coevolution: methodological and theoretical update. Syst Zool 39:14–30

Brooks DR (2004) Reticulations in historical biogeography: the triumph of time over space in evolution. In: Lomolino MV, Heaney LR (eds) Frontiers of biogeography: new directions in the geography of nature. Sinauer Associates, Sunderland, MA, pp 125–144

Brooks DR (2005) Historical biogeography in the age of complexity: expansion and integration. Rev Mex Biodivers 76:79–94

Brooks DR, McLennan DA (1991) Phylogeny, ecology and behavior: a research program in comparative biology. University of Chicago Press, Chicago

Brooks DR, van Veller MGP, McLennan DA (2001) How to do BPA, really. J Biogeogr 28:345–358

Brooks DR, Dowling PG, van Veller MGP, Hoberg EP (2003) Ending a decade of deception: a valiant failure, a not-so valiant failure, and a success story. Cladistics 20:32–46

Cano JM, Gurrea P (2003) La distribución de las zigenas (Lepidoptera, Zygaenidae) ibéricas: Una consecuencia del efecto península. Graellsia 59:273–285

Carpenter JM (1992) Incidit in Scyllam qui vult vitare Charybdim. Cladistics 8:100–102

Cecca F (2002) Palaeobiogeography of marine fossil invertebrates: concepts and methods. Taylor and Francis, London

Cecca F, Morrone JJ, Ebach MC (2011) Biogeographical convergence and time-slicing: concepts and methods in cladistic biogeography. In: Upchurch P, McGowan A, Slater C (eds) Palaeogeography and palaeobiogeography: biodiversity in space and time, Systematics Association Special Volume. Taylor & Francis, CRC Press, Boca Raton, FL, pp 1–12

Cooper RA (1989) New Zealand tectonostratigraphic terranes and panbiogeography. New Zealand J Zool 16:699–712

Corral-Rosas V, Morrone JJ (2017) Analyzing the assembly of cenocrons in the Mexican transition zone through a time-sliced cladistic biogeographic analysis. Aust Syst Bot 29:489–501

Cox CB, Moore PD (1998) Biogeography: an ecological and evolutionary approach. Blackwell Science, Oxford

Cracraft J (1975) Historical biogeography and Earth history: perspectives for a future synthesis. Ann Missouri Bot Gard 62:227–250
Cracraft J (1988) Deep-history biogeography: retrieving the historical pattern of evolving continental biotas. Syst Zool 37:221–236
Cracraft J (1991) Patterns of diversification within continental biotas: hierarchical congruence among the areas of endemism of Australian vertebrates. Aust Syst Bot 4:211–227
Craw RC (1988) Continuing the synthesis between panbiogeography, phylogenetic systematics and geology as illustrated by empirical studies on the biogeography of New Zealand and the Chatham Islands. Syst Zool 37:291–310
Craw RC, Page R (1988) Panbiogeography: method and metaphor in the new biogeography. In: Ho MW, Fox SW (eds) Evolutionary processes and metaphors. Wiley, Chichester, pp 163–189
Craw RC, Grehan JR, Heads MJ (1999) Panbiogeography: tracking the history of life. Oxford biogeography series 11, New York and Oxford
Crisci JV, Katinas L (2009) Darwin, historical biogeography, and the importance of overcoming binary oppositions. J Biogeogr 36:1027–1032
Crisci JV, Katinas L, Posadas P (2003) Historical biogeography: an introduction. Harvard University Press, Cambridge, MA
Crisp MD, Linder HP, Weston PH (1995) Cladistic biogeography of plants in Australia and New Guinea: congruent pattern revealed two endemic tropical tracks. Syst Zool 44:457–473
Croizat L (1958) Panbiogeography, vols 1 and 2. Published by the Author, Caracas
Croizat L (1964) Space, time, form: the biological synthesis. Published by the Author, Caracas
Croizat L (1982) Vicariance/vicariism, panbiogeography, "vicariance biogeography", etc: a clarification. Syst Zool 31:291–304
Croizat L, Nelson G, Rosen DE (1974) Centers of origin and related concepts. Syst Zool 23:265–287
Crother BI, Murray C (2011) Ontology of areas of endemism. J Biogeogr 38:1009–1015
Cué-Bär EM, Villaseñor JL, Morrone JJ, Ibarra-Manríquez G (2006) Identifying priority areas for conservation in Mexican tropical deciduous forest based on tree species. Interciencia 31:712–719
Darlington PJ Jr (1957) Zoogeography: the geographical distribution of animals. Wiley, New York
Darwin CR (1859) On the origin of species by means of natural selection, or the preservation of favoured races in the struggle for life. John Murray, London
DaSilva MB, Pinto-da-Rocha R, DeSouza AM (2015) A protocol for the delimitation of areas of endemism and the historical regionalization of the Brazilian Atlantic rain forest using harvestmen distribution data. Cladistics 31:692–705
de Queiroz A (2014) The monkey's voyage: how improbable journeys shaped the history of life. Basic Books, New York
Donoghue MJ, Moore BR (2003) Toward an integrative historical biogeography. Integr Comp Biol 43:261–270
Dowling APG (2002) Testing the accuracy of TreeMap and Brooks parsimony analyses of coevolutionary patterns using artificial associations. Cladistics 18:416–435
Ebach MC, Edgecombe GD (2001) Cladistic biogeography: component-based methods and paleontological application. In: Adrain JM, Edgecombe GD, Lieberman BS (eds) Fossils, phylogeny, and form: an analytical approach. Kluwer/Plenum, New York, pp 235–289
Ebach MC, Parenti LR (2015) The dichotomy of the modern bioregionalization revival. J Biogeogr 42:1801–1808
Ebach MC, Humphries CJ, Williams DM (2003) Phylogenetic biogeography deconstructed. J Biogeogr 30:1285–1296
Ebach MC, Humphries CJ, Newman RA, Williams DM, Walsh SA (2005) Assumption 2: opaque to intuition? J Biogeogr 32:781–787
Ebach MC, Morrone JJ, Parenti LR, Viloria AL (2008) International code of area nomenclature. J Biogeogr 35:1153–1157

Echeverría-Londoño S, Miranda-Esquivel DR (2011) MartiTracks: a geometrical approach for identifying geographical patterns of distribution. PLoS One 6:1–7
Echeverry A, Morrone JJ (2010) Parsimony analysis of endemicity as a panbiogeographical tool: an analysis of Caribbean plant taxa. Biol J Linn Soc 101:961–976
Echeverry A, Silva-Romo G, Morrone JJ (2012) Tectonostratigraphic terrane relationships: a glimpse into the Caribbean under a cladistic approach. Palaeogeogr Palaeoclimatol Palaeoecol 353–355:87–92
Enghoff H (1995) Historical biogeography of the Holarctic: area relationships, ancestral areas, and dispersal of non-marine animals. Cladistics 11:223–263
Enghoff H (1996) Widespread taxa, sympatry, dispersal, and an algorithm for resolved area cladograms. Cladistics 12:349–364
Enghoff H (2000) Reversals as branch support in biogeographical parsimony analysis. Vie Mil 50:255–260
Escalante T (2009) Un ensayo sobre regionalización biogeográfica. Rev Mex Biodivers 80:551–560
Escalante T (2015) Parsimony analysis of endemicity and analysis of endemicity: a fair comparison. Syst Biodivers 13:413–418
Escalante T (2017) A natural regionalization of the world based on primary biogeographic homology of terrestrial mammals. Biol J Linn Soc 120:349–362
Escalante T, Espinosa D, Morrone JJ (2003) Using parsimony analysis of endemicity to analyze the distribution of Mexican land mammals. Southwest Nat 48:563–578
Escalante T, Rodríguez G, Morrone JJ (2004) The diversification of the Nearctic mammals in the Mexican Transition Zone. Biol J Linn Soc 83:327–339
Escalante T, Noguera-Urbano EA, Pimentel B, Aguado-Bautista O (2017) Methodological issues in modern track analysis. Evol Biol 44:284–293
Felsenstein J (1986) Phylogenetic inference package (PHYLIP). University of Washington, Seatle
Ferrari A, Barão KR, Simões FL (2013) Quantitative panbiogeography: was the congruence problem solved? Syst Biodivers 11:285–302
Ferro I, Morrone JJ (2014) Biogeographic transition zones: a search for conceptual synthesis. Biol J Linn Soc 113:1–12
Folinsbee KE, Evans DC (2012) A protocol for temporal calibration of general area cladograms. J Biogeogr 39:688–697
Fontenla JL, López Admirall A (2008) Archipiélago cubano: biogeografía histórica y complejidad. CubaLibri, Washington, DC
Fortino AD, Morrone JJ (1997) Signos gráficos para la representación de análisis panbiogeográficos. Biogeographica 73:49–56
Gámez N, Nihei SS, Scheinvar E, Morrone JJ (2017) A temporally dynamic approach for cladistic biogeography and the processes underlying the biogeographic patterns of North American deserts. J Zool Syst Evol Res 55:11–18
García-Barros E (2003) Mariposas diurnas endémicas de la región Paleártica Occidental: Patrones de distribución y su análisis mediante parsimonia (Lepidoptera, Papilionoidea). Graellsia 59:233–258
García-Barros E, Gurrea P, Luciáñez MJ, Cano JM, Munguira ML, Moreno JC, Sainz H, Sanza MJ, Simón JC (2002) Parsimony analysis of endemicity and its application to animal and plant geographical distributions in the Ibero-Balearic region (western Mediterranean). J Biogeogr 29:109–124
García-Trejo EA, Navarro AG (2004) Patrones biogeográficos de la riqueza de especies y el endemismo de la avifauna en el oeste de México. Acta Zool Mex 20:167–185
Gillespie JH (1991) The causes of molecular evolution. Oxford University Press, New York
Goloboff P (2011) NDM/VNDM v. 3.0 Programs for identification of areas of endemism. Program and documentation. http://www.lillo.org.ar/phylogeny/
Goloboff PA, Farris JS, Nixon KC (2008) TNT, a free program for phylogenetic analysis. Cladistics 24:774–786

Goyenechea I, Flores Villela O, Morrone JJ (2001) Introducción a los fundamentos y métodos de la biogeografía cladística. In: Llorente Bousquets J, Morrone JJ (eds) Introducción a la biogeografía en Latinoamérica: conceptos, teorías, métodos y aplicaciones, Las Prensas de Ciencias. UNAM, Mexico City, pp 225–232

Grande L (1985) The use of paleontology in systematics and biogeography, and a time control refinement for historical biogeography. Paleobiology 11:234–243

Grehan JR (2001a) Islas Galápagos: Biogeografía, tectónica y evolución en un archipiélago oceánico. In: Llorente Bousquets J, Morrone JJ (eds) Introducción a la biogeografía en Latinoamérica: Conceptos, teorías, métodos y aplicaciones, Las Prensas de Ciencias. UNAM, Mexico City, pp 153–160

Grehan JR (2001b) Panbiogeografía y la geografía de la vida. In: Llorente Bousquets J, Morrone JJ (eds) Introducción a la biogeografía en Latinoamérica: conceptos, teorías, métodos y aplicaciones, Las Prensas de Ciencias. UNAM, Mexico City, pp 181–195

Halffter G (1987) Biogeography of the montane entomofauna of Mexico and Central America. Annu Rev Entomol 32:95–114

Halffter G, Morrone JJ (2017) An analytical review of Halffter's Mexican transition zone, and its relevance for evolutionary biogeography, ecology and biogeographical regionalization. Zootaxa 4226:1–46

Hallam A (1974) Changing patterns of provinciality and diversity of fossil animals in relation to plate tectonics. J Biogeogr 1:213–225

Harold AS, Mooi RD (1994) Areas of endemism: definition and recognition criteria. Syst Biol 43:261–266

Hausdorf B (2002) Units in biogeography. Syst Biol 51:648–652

Heads MJ (1989) Integrating earth and life sciences in New Zealand natural history: the parallel arcs model. New Zealand J Zool 16:549–585

Heads MJ (2004) What is a node? J Biogeogr 31:1883–1891

Heads MJ (2005) The history and philosophy of panbiogeography. In: Llorente Bousquets J, Morrone JJ (eds) Regionalización biogeográfica en Iberoamérica y tópicos afines: Primeras Jornadas Biogeográficas de la Red Iberoamericana de Biogeografía y Entomología Sistemática (RIBES XII.I-CYTED), Las Prensas de Ciencias. UNAM, Mexico City, pp 67–123

Heads M (2009) Inferring biogeographic history from molecular phylogenies. Biol J Linn Soc 98:757–774

Heads M (2011) Old taxa on young islands: a critique of the use of island age to date island endemic clades and calibrate phylogenies. Syst Biol 60:204–218

Henderson IM (1989) Quantitative panbiogeography: an investigation into concepts and methods. New Zealand J Zool 16:495–510

Hennig W (1950) Grundzüge einer Theorie der phylogenetischen Systematik. Deutscher Zentralverlag, Berlin

Holt BG, Lessard JP, Borregaard MK, Fritz SA, Araújo MB, Dimitrov D, Fabre PH, Graham CH, Graves GR, Jønsson KA, Nogués-Bravo D, Wang Z, Whittaker RJ, Fjeldså RJ, Rahbek C (2013) An update of Wallace's zoogeographic regions of the world. Science 339:74–78

Hovenkamp P (1997) Vicariance events, not areas, should be used in biogeographic analysis. Cladistics 13:67–79

Hovenkamp P (2001) A direct method for the analysis of vicariance patterns. Cladistics 17:260–265

Hull DL (1983) Popper and Plato's metaphor. In: Platnick NI, Funk VA (eds) Advances in cladistics, Proceedings of the second meeting of the Willi Hennig Society, vol 2. Columbia University Press, New York, pp 177–189

Humphries CJ (1989) Any advance on assumption 2? J Biogeogr 16:101–102

Humphries CJ (1992) Cladistic biogeography. In: Forey PL, Humphries CJ, Kitching IJ, Scotlandm RW, Siebert DJ, Williams DM (eds) Cladistics: a practical course in systematics, The Systematics Association Publication 10. Oxford Science Publications, Clarendon Press, Oxford, pp 137–159

Humphries CJ (2000) Form, space and time; which comes first? J Biogeogr 27:11–15

Humphries CJ, Ebach MC (2004) Biogeography on a dynamic Earth. In: Lomolino MV, Heaney LR (eds) Frontiers of biogeography: new directions in the geography of nature. Sinauer Associates, Sunderland, MA, pp 67–86
Humphries CJ, Parenti LR (1999) Cladistic biogeography: Interpreting patterns of plant and animal distributions, 2nd edn. Oxford University Press, Oxford
Hunn CA, Upchurch P (2001) The importance of time/space in diagnosing the causality of phylogenetic events: towards a "chronobiogeographical paradigm". Syst Biol 50:391–407
Kamp PJJ (1980) Pacifica and New Zealand, proposed eastern elements in Gondwanaland's history. Nature 288:659–664
King AR, Ebach M (2017) A novel approach to time-slicing areas within biogeographic-area classifications: Wallacea as an example. Aust Syst Bot 30:495–512
Kishino H, Thorne JL, Bruno WJ (2001) Performance of a divergence time estimation method under a probabilistic model of rate evolution. Mol Biol Evol 18:352–361
Kluge AG (1988) Parsimony in vicariance biogeography: a quantitative method and a greater Antillean example. Syst Zool 37:315–328
Knowles LL, Maddison WP (2002) Statistical phylogeography. Mol Ecol 11:2623–2635
Kreft H, Jetz W (2010) A framework for delineating biogeographical regions based on species distributions. J Biogeogr 37:2029–2053
Kuch M, Rohland N, Betancourt JL, Latorre CL, Steppan S, Poinar HN (2002) Molecular analysis of an 11,700 year old rodent midden from the Atacama desert, Chile. Mol Ecol 11:913–924
Lanteri AA, Confalonieri VA (2003) Filogeografía: objetivos, métodos y ejemplos. In: Morrone JJ, Llorente Bousquets J (eds) Una perspectiva latinoamericana de la biogeografía, Las Prensas de Ciencias. UNAM, Mexico City, pp 185–193
León-Paniagua L, García E, Arroyo-Cabrales J, Castañeda-Rico S (2004) Patrones biogeográficos de la mastofauna. In: Luna I, Morrone JJ, Espinosa D (eds) Biodiversidad de la Sierra Madre Oriental, Las Prensas de Ciencias. UNAM, Mexico City, pp 469–486
Lieberman BS (1997) Early Cambrian paleogeography and tectonic history: a biogeographic approach. Geology 25:1039–1042
Lieberman BS (2000) Paleobiogeography: using fossils to study global change, plate tectonics and evolution. Kluwer Academic, New York
Lieberman BS (2003) Paleobiogeography: the relevance of fossils to biogeography. Annu Rev Ecol Syst 34:51–69
Lieberman BS (2004) Range expansion, extinction, and biogeographic congruence: a deep time perspective. In: Lomolino MV, Heaney LR (eds) Frontiers of biogeography: new directions in the geography of nature. Sinauer Associates, Sunderland, MA, pp 111–124
Linder HP (2001) On areas of endemism, with an example of the African Restionaceae. Syst Biol 50:892–912
Linder HP, Mann DM (1998) The phylogeny and biogeography of *Thamnochortus* (Restionaceae). Bot J Linn Soc London 128:319–357
Lobo JM (1999) Individualismo y adaptación espacial: Un nuevo enfoque para explicar la distribución geográfica de las especies. Bol Soc Entomol Aragon 26:561–572
Lobo JM (2007) Los "patrones de dispersión" de la fauna ibérica de Scarabaeinae (Coleoptera). In: Zunino M, Melic A (eds) Escarabajos, diversidad y conservación biológica: Ensayos en homenaje a Gonzalo Halffter, Sociedad Entomológica Aragonesa, Monografías 3er. Milenio M3M, Zaragoza, pp 159–177
Lomolino MV, Riddle BR, Brown JH (2006) Biogeography, 3rd edn. Sinauer Associates, Sunderland, MA
Luna-Vega I, Alcántara O, Morrone JJ, Espinosa Organista D (2000) Track analysis and conservation priorities in the cloud forests of Hidalgo, Mexico. Divers Distrib 6:137–143
MacDonald GM (2003) Biogeography: space, time, and life. Wiley, New York
Magallón SA (2004) Dating lineages: molecular and paleontological approaches to the temporal framework of clades. Int J Plant Sci 165:S7–S21

Maldonado M, Uriz MJ (1995) Biotic affinities in a transitional zone between the Atlantic and the Mediterranean: a biogeographical approach based on sponges. J Biogeogr 22:89–110
McCarthy D (2003) The trans-Pacific zipper effect: disjunct sister taxa and matching geological outlines that link the Pacific margins. J Biogeogr 30:1545–1561
McCarthy D (2007) Are plate tectonic explanations for trans-Pacific disjunctions plausible? Empirical tests of radical dispersalist theories. In: Ebach MC, Tangney RS (eds) Biogeography in a changing world, The Systematics Association Special Volume Series 70. CRC Press, Boca Raton, FL, pp 177–198
McKenna MC (1973) Sweeptakes, filters, corridors, Noah's ark and beached Viking funeral ships in paleogeography. In: Tarling DH, Runcorn SK (eds) Implications of continental drift to the earth sciences, vol I. Academic, London, pp 295–308
Michaux B (1989) Generalized tracks and geology. Syst Zool 38:390–398
Miguel-Talonia C, Escalante T (2013) Los nodos: El aporte de la panbiogeografía al entendimiento de la biodiversidad. Biogeografía 6:30–42
Miranda Esquivel DR, Donato M, Posadas P (2003) La dispersión ha muerto, larga vida a la dispersión. In: Morrone JJ, Llorente Bousquets J (eds) Una perspectiva latinoamericana de la biogeografía, Las Prensas de Ciencias. UNAM, Mexico City, pp 179–184
Mitchell SD (2002) Integrative pluralism. Biol Philos 17:55–70
Mitchell SD, Dietrich MR (2006) Integration without unification: an argument for pluralism in the biological sciences. Am Nat 168:573–579
Morrone JJ (1994) On the identification of areas of endemism. Syst Biol 43:438–441
Morrone JJ (2001) Homology, biogeography and areas of endemism. Divers Distrib 7:297–300
Morrone JJ (2004a) Homología biogeográfica: Las coordenadas espaciales de la vida. Cuadernos del Instituto de Biología 37, Instituto de Biología, UNAM, Mexico City
Morrone JJ (2004b) Panbiogeografía, componentes bióticos y zonas de transición. Rev Bras Entomol 48:149–162
Morrone JJ (2006) Biogeographic areas and transition zones of Latin America and the Caribbean Islands based on panbiogeographic and cladistic analyses of the entomofauna. Annu Rev Entomol 51:467–494
Morrone JJ (2008) Endemism. In: Jorgensen SE, Fath BD (eds) Encyclopedia of ecology. Elsevier B. V., Oxford, pp 1254–1259
Morrone JJ (2009) Evolutionary biogeography: an integrative approach with case studies. Columbia University Press, New York
Morrone JJ (2014a) On biotas and their names. Syst Biodivers 12:386–392
Morrone JJ (2014b) Parsimony analysis of endemicity (PAE) revisited. J Biogeogr 41:842–854
Morrone JJ (2014c) Cladistic biogeography of the Neotropical region: Identifying the main events in the diversification of the terrestrial biota. Cladistics 30:202–214
Morrone JJ (2015) Track analysis beyond panbiogeography. J Biogeogr 42:413–425
Morrone JJ (2017) Neotropical biogeography: regionalization and evolution. CRC Press, Taylor and Francis Group, Boca Raton, FL
Morrone JJ (2018) The spectre of biogeographic regionalization. J Biogeogr 45:282–288
Morrone JJ, Carpenter JM (1994) In search of a method for cladistic biogeography: an empirical comparison of component analysis, Brooks parsimony analysis, and three-area statements. Cladistics 10:99–153
Morrone JJ, Crisci JV (1995) Historical biogeography: introduction to methods. Annu Rev Ecol Syst 26:373–401
Morrone JJ, Escalante T (2002) Parsimony analysis of endemicity (PAE) of Mexican terrestrial mammals at different area units: when size matters. J Biogeogr 29:1095–1104
Morrone JJ, Márquez J (2001) Halffter's Mexican Transition Zone, beetle generalised tracks, and geographical homology. J Biogeogr 28:635–650
Morrone JJ, Roig-Juñent S, Crisci JV (1994) Cladistic biogeography of terrestrial subantarctic beetles (Insecta: Coleoptera) from South America. Nat Geogr Res Expl 10:104–115

Müller P (1973) The dispersal centres of terrestrial vertebrates in the Neotropical realm: a study in the evolution of the Neotropical biota and its native landscapes. Junk, The Hague
Müller P (1979) Introducción a la zoogeografía. Blume, Barcelona
Murray CM, Crother BI (2016) Entities on a temporal scale. Acta Biotheor 64:1–10
Myers AA (1991) How did Hawaii accumulate its biota?: A test from the Amphipoda. Global Ecol Biogeogr Lett 1:24–29
Navarro AG, Garza-Torres HA, López de Aquino S, Rojas-Soto OR, Sánchez-González LA (2004) Patrones biogeográficos de la avifauna. In: Luna I, Morrone JJ, Espinosa D (eds) Biodiversidad de la Sierra Madre Oriental, Las Prensas de Ciencias. UNAM, Mexico City, pp 439–467
Nelson G (1984) Cladistics and biogeography. In: Duncan T, Stuessy TF (eds) Cladistics: perspectives on the reconstruction of evolutionary history. Columbia University Press, New York, pp 273–293
Nelson G, Ladiges PY (1996) Paralogy in cladistic biogeography and analysis of paralogy-free subtrees. Am Mus Novit 3167:1–58
Nelson G, Ladiges PY (2001) Gondwana, vicariance biogeography and the New York school revisited. Aust J Bot 49:389–409
Nelson G, Ladiges PY (2003) Geographic paralogy. In: Morrone JJ, Llorente Bousquets J (eds) Una perspectiva latinoamericana de la biogeografía, Las Prensas de Ciencias. UNAM, Mexico City, pp 173–177
Nelson G, Platnick NI (1978) The perils of plesiomorphy: widespread taxa, dispersal, and phenetic biogeography. Syst Zool 27:474–477
Nelson G, Platnick NI (1981) Systematics and biogeography: cladistics and vicariance. Columbia University Press, New York
Nihei SS (2006) Misconceptions about parsimony analysis of endemicity. J Biogeogr 33:2099–2106
Nihei SS, de Carvalho CJB (2005) Distributional patterns of the Neotropical fly genus *Polietina* Schnabl and Dziedzicki (Diptera: Muscidae); A phylogeny-supported analysis using panbiogeographic tools. Pap Avuls Zool 45:313–326
Noguera-Urbano EA (2016) Areas of endemism: travelling through space and the unexplored dimension. Syst Biodivers 14:131–139
Nur A, Ben-Avraham Z (1980) Lost Pacifica continent: a mobilistic speculation. In: Rosen DE, Nelson G (eds) Vicariance biogeography: a critique. Columbia University Press, New York, pp 341–358
Oliveira U, Brescovit AD, Santos AJ (2015) Delimiting areas of endemism through kernel interpolation. PLoS One 10:e0116673
Olivero J, Real R, Vargas JM (1998) Distribution of breeding, wintering, and resident waterbirds in Europe: biotic regions and the macroclimate. Ornis Fenn 75:153–175
Olivero J, Real R, Márquez AL (2011) Fuzzy chorotypes as a conceptual tool to improve insight into biogeographic patterns. Syst Biol 60:645–660
Page RDM (1987) Graphs and generalized tracks: quantifying Croizat's panbiogeography. Syst Zool 36:1–17
Page RDM (1988) Quantitative cladistic biogeography: constructing and comparing area cladograms. Syst Zool 37:254–270
Page RDM (1989) Comments on component-compatibility in historical biogeography. New Zealand J Zool 16:471–483
Page RDM (1990) Component analysis: a valiant failure? Cladistics 6:119–136
Page RDM (1993) Component user's manual. Release 2.0. The Natural History Museum, London
Page RDM, Holmes EC (1998) Molecular evolution: a phylogenetic approach. Blackwell Science, Oxford
Parenti LR (1981) Discussion. In: Nelson G, Rosen DE (eds) Vicariance biogeography: a critique. Columbia University Press, New York, pp 490–497
Parenti LR (2007) Common cause and historical biogeography. In: Ebach MC, Tangney RS (eds) Biogeography in a changing world, The Systematics Association Special Volume Series 70. CRC Press, Boca Raton, FL, pp 61–82

Parenti LR, Ebach MC (2009) Comparative biogeography: discovering and classifying biogeographical patterns of a dynamic Earth. University of California Press, Berkeley

Platnick NI, Nelson G (1978) A method of analysis for historical biogeography. Syst Zool 27:1–16

Popper KR (1959) The logic of scientific discovery. Hutchinson, London

Popper KR (1963) Conjectures and refutations: the growth of scientific knowledge. Routledge, London

Porzecanski AL, Cracraft J (2005) Cladistic analysis of distributions and endemism (CADE): using raw distributions of birds to unravel the biogeography of the South American aridlands. J Biogeogr 32:261–275

Rambaut A, Bromham L (1998) Estimating divergence dates from molecular sequences. Mol Biol Evol 15:442–448

Rapoport EH (1975) Areografía: Estrategias geográficas de las especies. Fondo de Cultura Económica, Mexico City

Reig OA (1981) Teoría del origen y desarrollo de la fauna de mamíferos de América del Sur. Museo Municipal de Ciencias Naturales Lorenzo Scaglia, Mar del Plata

Ribichich AM (2005) From null community to non-randomly structured actual plant assemblages: parsimony analysis of species co-ocurrences. Ecography 28:88–98

Riddle BR, Hafner DJ (2004) The past and future roles of phylogeography in historical biogeography. In: Lomolino MV, Heaney LR (eds) Frontiers of biogeography: new directions in the geography of nature. Sinauer Associates, Sunderland, MA, pp 93–110

Riddle BR, Hafner DJ (2006) A step-wise approach to integrating phylogeographic and phylogenetic biogeographic perspectives on the history of a core North American warm deserts biota. J Arid Environ 66:435–461

Riddle BR, Dawson MN, Hadly EA, Hafner DJ, Hickerson MJ, Mantooth SJ, Yoder AD (2008) The role of molecular genetics in sculpting the future of integrative biogeography. Progr Phys Geogr 32:173–202

Rieppel O (1991) Things, taxa and relationships. Cladistics 7:93–100

Rieppel O (2004) The language of systematics, and the philosophy of 'total evidence'. Syst Biodivers 2:9–19

Roig-Juñent S (1994) Historia biogeográfica de América del Sur austral. Multequina (Mendoza) 3:167–203

Roig-Juñent S, Crisci JV, Posadas P, Lagos S (2002) Áreas de distribución y endemismo en zonas continentals. In: Costa C, Vanin SA, Lobo JM, Melic A (eds) Proyecto de Red Iberoamericana de Biogeografía y Entomología Sistemática PrIBES 2002, Monografías Tercer Milenio, vol 2. Sociedad Entomológica Aragonesa, Zaragoza, pp 247–266

Roig-Juñent SA, Griotti M, Domínguez MC, Agrain FA, Campos-Soldini P, Carrara R, Cheli G, Fernández-Campón F, Flores GE, Katinas L, Muzón JR, Neita-Moreno JC, Pessacq P, San Blas G, Scheibler EE, Crisci JV (2018) The Patagonian Steppe biogeographic province: Andean region or South American transition zone? Zool Scripta 47:623–629

Ronquist F, Sanmartín I (2011) Phylogenetic methods in biogeography. Annu Rev Ecol Evol Syst 42:441–464

Rosen BR (1985) Long-term geographical controls on regional diversity. J Open Univ Geol Soc 6:25–30

Rosen BR (1988) From fossils to earth history: applied historical biogeography. In: Myers AA, Giller PS (eds) Analytical biogeography: an integrated approach to the study of animal and plant distributions. Chapman and Hall, London, pp 437–481

Rosen DE, Nelson G (eds) (1980) Vicariance biogeography: a critique. Columbia University Press, New York

Rosen BR, Smith AB (1988) Tectonics from fossils?: Analysis of reef-coral and sea-urchin distributions from late cretaceous to recent, using a new method. In: Audley-Charles MG, Hallam A (eds.) Gondwana and Tethys, Geological Society Special Publication nr. 37, London, pp 275–306

Rutschmann F (2006) Molecular dating of phylogenetic trees: a brief review of current methods that estimate divergence times. Divers Distrib 12:35–48
Sanderson MJ (1998) Estimating rate and time in molecular phylogenies: beyond the molecular clock? In: Soltis DE, Soltis PS, Doyle JJ (eds) Molecular systematics of plants II: DNA sequencing. Kluwer Academic, Boston, pp 242–264
Sanderson MJ (2002) Estimating absolute rates of molecular evolution and divergence times: a penalized likelihood approach. Mol Biol Evol 19:101–109
Sanmartín I, Ronquist F (2002) New solutions to old problems: widespread taxa, redundant distributions and missing areas in event-based biogeography. Anim Biodivers Conserv 25:75–93
Sanmartín I, Ronquist F (2004) Southern hemisphere biogeography inferred by event-based models: plant versus animal patterns. Syst Biol 53:216–243
Santos CMD (2005) Parsimony analysis of endemicity: time for an epitaph? J Biogeogr 32:1284–1286
Santos CMD, Amorim DS (2007) Why biogeographical hypotheses need a well supported phylogenetic framework: a conceptual evaluation. Pap Avuls Zool 47:63–73
Santos CMD, Capellari RS (2009) On reciprocal illumination and consilience in biogeography. Evol Biol 36:407–415
Savage JM (1982) The enigma of the Central American herpetofauna: dispersals or vicariance? Ann Missouri Bot Gard 69:464–547
Sclater PL (1858) On the general geographic distribution of the members of the class Aves. Proc Linn Soc London Zool 2:130–145
Seberg O (1991) Biogeographic congruence in the South Pacific. Aust Syst Bot 4:127–136
Shields O (1979) Evidence for the initial opening of the Pacific Ocean in the Jurassic. Palaeogeogr Palaeoclimatol Palaeoecol 26:181–220
Shields O (1991) Pacific biogeography and rapid Earth expansion. J Biogeogr 18:583–585
Shields O (1996) Plate tectonics or an expanding Earth? J Geol Soc India 47:399–408
Siddall ME (2005) Bracing for another decade of deception: the promise of secondary Brooks parsimony analysis. Cladistics 21:90–99
Siddall ME, Perkins SL (2003) Brooks parsimony analysis: a valiant failure. Cladistics 19:554–564
Simpson GG (1953) Evolution and geography: an essay on historical biogeography with special reference to mammals. Condon Lecture Series, Oregon State System of Higher Education, Eugene
Simpson GG (1980) Splendid isolation: the curious history of South American mammals. Yale University Press, New Haven
Soltis DE, Soltis PS, Milligan BG (1992) Intraspecific chloroplast DNA variation: systematic and phylogenetic implications. In: Soltis PS, Soltis DE, Doyle JJ (eds) Molecular systematics of plants. Chapman and Hall, New York, pp 117–150
Swenson U, Backlund A, McLoughlin S, Hill RS (2001) *Nothofagus* biogeography revisited with special emphasis on the enigmatic distribution of subgenus *Brassospora* in New Caledonia. Cladistics 17:28–47
Swenson U, Nylinder S, Wagstaff S (2012) Are Asteraceae 1.5 billion years old? A reply to heads. Syst Biol 61:522–533
Szumik CA, Cuezzo F, Goloboff PA, Chalup AE (2002) An optimality criterion to determine areas of endemism. Syst Biol 51:806–816
Szumik CA, Casagranda D, Roig-Juñent S (2006) Manual de NDM/VNDM: programas para la identificación de areas de endemismo. Inst Argent Est Filog 5:1–26
Templeton AR (2004) Statistical phylogeography: methods of evaluating and minimizing inference errors. Mol Ecol 13:789–809
Trejo-Torres JC (2003) Biogeografía ecológica de las Antillas: Ejemplos de las orquídeas y las selvas cársticas. In: Morrone JJ, Llorente Bousquets J (eds) Una perspectiva latinoamericana de la biogeografía, Las Prensas de Ciencias. UNAM, Mexico City, pp 199–208
Trejo-Torres JC, Ackerman JD (2001) Biogeography of the Antilles based on a parsimony analysis of orchid distributions. J Biogeogr 28:775–794

Trejo-Torres JC, Ackerman JD (2002) Composition patterns of Caribbean limestone forests: Are parsimony, classification, and ordination analyses congruent? Biotropica 34:502–515

Upchurch P, Hunn CA (2002) "Time": The neglected dimension in cladistic biogeography? Geobios 35:277–286

Upchurch P, Hunn CA, Norman DB (2002) An analysis of dinosaurian biogeography: evidence for the existence of vicariance and dispersal patterns caused by geological events. Proc Roy Soc London Ser B 269:613–621

van Veller MGP, Zandee M, Kornet DJ (1999) Two requirements for obtaining valid common patterns under different assumptions in vicariance biogeography. Cladistics 15:393–406

van Veller MGP, Kornet DJ, Zandee M (2000) Methods in vicariance biogeography: assessment of the implementations of assumptions 0, 1, and 2. Cladistics 16:319–345

van Welzen PC, Turner H, Roos MC (2001) New Guinea: a correlation between accreting areas and dispersing Sapindaceae. Cladistics 17:242–247

Vargas JM (1992) Escuelas y tendencias en biogeografía histórica. Monogr Herpetol 2:107–136

Vilhena DA, Antonelli A (2015) A network approach for identifying and delimiting biogeographical regions. Nat Commun 6:1–9

Wallace AR (1876) The geographical distribution of animals, with a study of the relations of living and extinct faunas as elucidating the past changes of the Earth's surface. Macmillan, London

Wegener A (1912) Die Entstehung der Kontinente. Geol Rundsch 3:276–292

Wegener A (1929) The origin of continents and oceans. Dover, New York

Wiley EO (1981) Phylogenetics: the theory and practice of phylogenetic systematics. Wiley-Interscience, New York

Wiley EO (1987) Methods in vicariance biogeography. In: Hovenkamp P, Gittenberger E, Hennipman E, de Jong R, Roos MC, Sluys R, Zandee M (eds) Systematics and evolution: a matter of diversity. Utrecht, Institute of Systematic Botany, Utrecht University, pp 283–306

Wiley EO (1988a) Parsimony analysis and vicariance biogeography. Syst Zool 37:271–290

Wiley EO (1988b) Vicariance biogeography. Annu Rev Ecol Syst 19:513–542

Young GC (1995) Application of cladistics to terrane history – parsimony analysis of qualitative geological data. J SE Asian Earth Sci 11:167–176

Zandee M, Roos MC (1987) Component-compatibility in historical biogeography. Cladistics 3:305–332

Zuckerland E, Pauling L (1965) Molecular disease, evolution and genetic heterogeneity. In: Kasha M, Pullman B (eds) Horizons in biochemistry. Academic, London, pp 189–225

Zunino M, Zullini A (1995) Biogeografia: La dimensione spaziale dell'evoluzione. Casa Editrice Ambrosiana, Milan

Zunino M, Barbero E, Palestrini C, Buffa E, Roggero A (1996) Ipotesi biogeografiche versus ipotesi filogenetiche: Il genere *Typhaeus* Leach (Coleoptera: Geotrupidae) e il popolamento dell' area Sarda. Biogeographia 18:455–476

Chapter 3
A Historical Perspective of the Mexican Transition Zone

History, Stephen said, is a nightmare from which I am trying to awake.

James Joyce (1922), Ulyses

Abstract Several authors have considered that the complex area where the Neotropical and Nearctic biotas overlap corresponds to a transition zone. In the strict sense that is followed in this book, the Mexican Transition Zone includes the highlands of Mexico, Guatemala, Honduras, El Salvador, and Nicaragua north of Lake Nicaragua, whereas northern Mexico, the United States and Canada belong to the Nearctic region, and the lowlands of the Pacific coast and the Gulf of Mexico, the Yucatán Peninsula, and Central America belong to the Neotropical region. In a series of contributions, Gonzalo Halffter provided a coherent theory that explains how cenocrons that evolved in different geographic areas assembled in the Mexican Transition Zone. I review herein the historical development of Halffter's theory, including the characterization of the dispersal or distributional patterns recognized by this author. These distributional patterns are considered to represent cenocrons, namely, sets of taxa that share the same biogeographic history and constitute identifiable subsets within a biota by their common biotic origin and evolutionary history. The biotic assembly of the Mexican Transition Zone is summarized into five stages, from the Jurassic-Cretaceous to the Pleistocene.

3.1 Introduction

Half a century ago, Gonzalo Halffter started a series of contributions devoted to the biogeographic analysis of Mexican Scarabaeidae (Halffter 1962, 1964a, b, 1965, 1968, 1972, 1974, 1975, 1976, 1978, 1987, 2003, 2017; Halffter et al. 1995). The most important concept of these contributions is the recognition of the Mexican Transition Zone, which Halffter defined as:

J. J. Morrone, *The Mexican Transition Zone*, https://doi.org/10.1007/978-3-030-47917-6_3

> a complex and varied area in which the Neotropical and the Nearctic faunas overlap. This area includes part of the southwestern United States, all of Mexico, and a large part of Central America extending to the Nicaraguan lowlands. The isthmus south of Lake Nicaragua (Costa Rica and Panama), although rich in endemisms, presents marked South American affinities and a Nearctic penetration similar to that of the northern part of South America. For these reasons I believe it is preferable not to consider this area as part of the Transition Zone. (Halffter 1987, p. 95)

The earliest reference to the Mexican Transition Zone may be traced to Wallace (1876). Wallace (1876) described and characterized the Mexican subregion as the northernmost part of the Neotropical region, which he considered to be transitional between the Neotropical and Nearctic regions. Heilprin (1887) followed Wallace's treatment of the Mexican subregion. Dugès (1902) compared the bird faunas of the United States and Mexico, considering that they were quite similar, but the Mexican Plateau was transitional between North and South America. Regarding the herpetofauna, he considered that the Mexican and Central American faunas were also transitional between North and South America. Hoffmann (1936) noted the presence of both Nearctic (septentrional) and Neotropical (meridional) insects in the Mexican fauna. Vivó (1943) identified a phytogeographic transition zone in Mesoamerica. Darlington (1957) formally named this area as the "Central American-Mexican Transition Zone," noting that "[i]t has received nothing like the attention given to Wallace's Line and Wallacea, although it is probably equally important" (Darlington 1957, p. 460). Rapoport (1971) referred to the line separating the Nearctic and Neotropical regions as the "Anáhuac line." Cabrera and Willink (1973) considered that the flora and fauna of this area were transitional between the Nearctic and Neotropical regions. Rzedowski (1978) recognized a Mountain Mesoamerican region, which he considered that: "does not belong definitely to the Holarctic or Neotropical kingdoms, because it has elements of both in important proportions" (Rzedowski 1978, p. 101). Ortega and Arita (1998) analyzed quantitatively the extension of the Mexican Transition Zone, based on species of Chiroptera.

Gonzalo Halffter (Fig. 3.1) has contributed more than any other author to the knowledge of the Mexican Transition Zone (Morrone 2005, 2010, 2015a). In a series of contributions on the systematics and biogeography of Mexican Scarabaeidae (Halffter 1962, 1964a, b, 1965, 1968, 1972, 1974, 1975, 1976, 1978, 1987, 2003; Reyes-Castillo and Halffter 1978; Kohlmann and Halffter 1988, 1990; Zunino and Halffter 1988; Halffter et al. 1995, 2008a, b, 2019; Lobo and Halffter 2000; Gutiérrez-Velázquez et al. 2013; Halffter and Morrone 2017), he provided a coherent theory that explains how sets of taxa that evolved in different geographic areas (cenocrons) assembled in the transition zone. This theory developed gradually, during half a century, following ideas initially formulated by Halffter (1962) that were refined and clarified in successive contributions from him and other authors (see historical reviews by Llorente Bousquets 1996; Reyes-Castillo 2003; Morrone 2015a; Halffter and Morrone 2017).

Fig. 3.1 Mexican entomologist and biogeographer Gonzalo Halffter (born 1932 in Madrid, Spain). (Photograph provided by Federico Escobar)

3.2 Halffter's Initial Contributions

In his first biogeographic contribution, Halffter (1962) described three dispersal patterns to account for the distributional patterns shown by species of Scarabaeoidea (Coleoptera) from the Mexican Transition Zone. He stated that his main objective was:

> to determine the patterns to which the dispersal of Mexican Scarabaeidae adjust and to compare these patterns with the characteristics of the two regions that enter into contact in our country (Halffter 1962, p. 1).

Halffter (1962) postulated that Scarabaeidae and other insects showed some discrepancies with the vertebrates used to characterize the classic zoogeographic regions, especially in the Mexican Plateau. In order to compare vertebrate and insect taxa, Halffter considered it is important to distinguish between a zoogeographic region (a geographically limited area, usually characterized by its vertebrates) and a fauna (the set of animal taxa with a center of radiation and a common geographic history). He considered that a zoogeographic region refers only to the present time, whereas a fauna implies the evolutionary and biogeographic development of a group and the conditions that have allowed it. In insects, and in contrast to vertebrates, a region and its fauna may not coincide. For example, Neotropical faunistic elements are present in the Nearctic region. In the case of the Mexican insects, Halffter found that this distinction was particularly relevant. Based on the geographic distribution of the Mexican species of Scarabaeidae-Laparosticti (currently subfamilies Scarabaeinae and Geotrupinae), Halffter (1962) characterized three "elements" or "dispersal patterns."

The Typical Neotropical dispersal pattern is comprised of faunistic elements that adjust to the boundaries of the Neotropical zoogeographic region, represented by genera widespread in South America, where they originally evolved. Some of the Mexican species may extend to Panama, reaching even northern South America. Ecologically, Typical Neotropical species are found in tropical forests of the lowlands and do not expand into the Mexican Plateau because of the contrasting xeric ecological conditions that prevail in it. Along the coasts of the Gulf of Mexico and the Pacific Ocean, Typical Neotropical species extend their distribution to the north, even reaching southern United States. The Scarabaeid *Canthon indigaceus* (Fig. 3.2), with three subspecies, was chosen by Halffter to illustrate this pattern. This species belong to a South American genus, with species distributed in South America, Central America, and the Mexican tropical lowlands. Halffter (1962) considered that Typical Neotropical elements may show different degrees of "penetration" in Mexico, from genera with several species in South America and few species in the country, that do not extend west of the isthmus of Tehuantepec or Chiapas

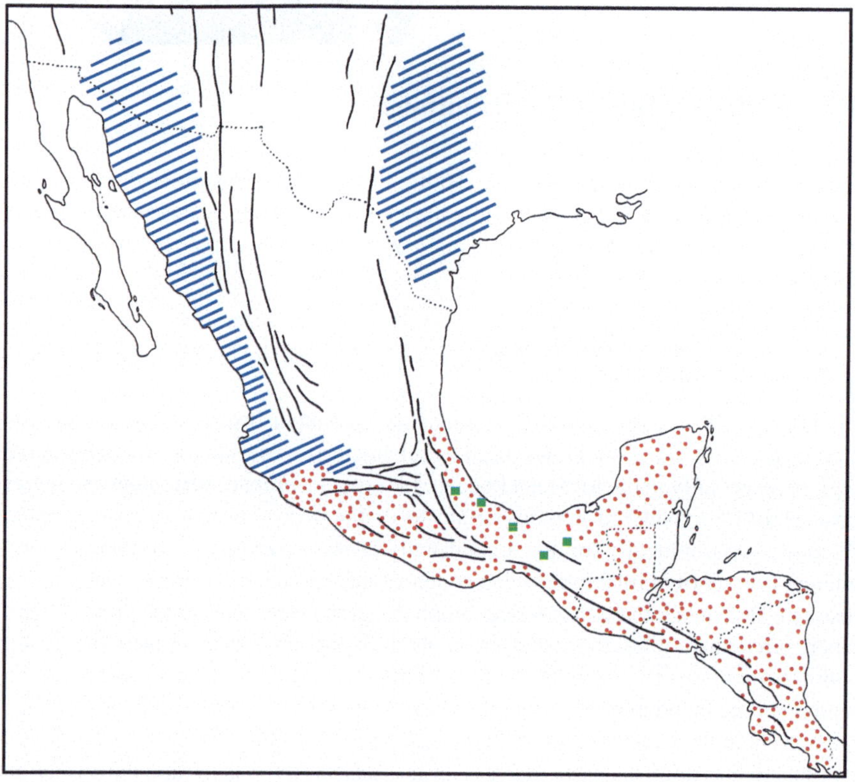

Fig. 3.2 Distribution of three subspecies of *Canthon indigaceus*, showing the Typical Neotropical dispersal pattern (modified from Halffter 1962). Blue lines, *Canthon indigaceus indigaceus*; red circles, *C. indigaceus chevrolati*; green squares, *C. indigaceus chiapas*

(e.g., *Uroxys*), to genera with more than one species in the country, some of which may even colonize the Mexican Plateau (e.g., *Ateuchus*).

The Mexican Plateau dispersal pattern includes Neotropical taxa that dispersed into the Mexican Plateau, geographically representing a "peninsula of the Nearctic region" (Halffter 1962, p. 6), from South America. Species inhabiting the Mexican Plateau are different to those of the previous element, although the genera to which they belong may be the same, and they may eventually disperse north of Mexico into southern United States, especially Arizona, Texas, Louisiana, and Florida. According to Halffter (1962), a typical species is *Canthon humectus* (Fig. 3.3), with six subspecies. Species of Scarabaeidae assigned to this dispersal pattern constitute more than 80% of the species in the Mexican Plateau. These taxa include the genera *Canthon*, *Phanaeus*, *Dichotomius*, *Boreocanthon*, and *Melanocanthon*. This abundance of Neotropical species of insects contrasts clearly with the characteristic vertebrate species of the Plateau, which are mostly Nearctic.

The Nearctic dispersal pattern includes North American taxa distributed in the Sierra Madre Occidental, Sierra Madre Oriental, Sierra Madre del Sur, Transmexican Volcanic Belt, Chiapas Highlands, and mountainous areas of Central America.

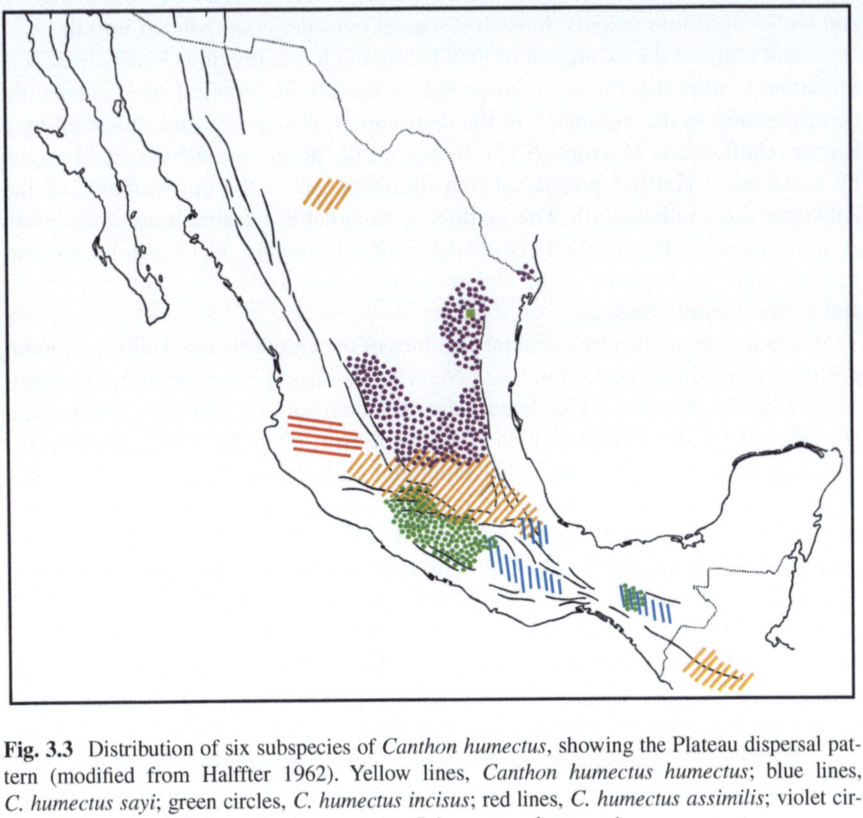

Fig. 3.3 Distribution of six subspecies of *Canthon humectus*, showing the Plateau dispersal pattern (modified from Halffter 1962). Yellow lines, *Canthon humectus humectus*; blue lines, *C. humectus sayi*; green circles, *C. humectus incisus*; red lines, *C. humectus assimilis*; violet circles, *C. humectus hidalgoensis*; green square, *C. humectus alvarengai*

Some species may have some scattered localities in the Mexican Plateau and the Balsas Basin. Species of Scarabaeidae assigned to the Nearctic dispersal pattern constitute 95–100% of the species in high mountain areas. Nearctic elements include the genera *Ceratotrupes* and *Geotrupes*.

Almost all the species of Scarabaeidae analyzed by Halffter (1962) could be assigned to one of these three patterns. The only exceptions are some species of *Sisyphus*, *Copris*, and *Onthophagus*, which, although distributed in the tropical lowlands, do not represent Neotropical taxa.

Halffter (1964a) published an essay on the origin and distribution of the entomofauna of the Americas (Halffter 1964b is a preliminary version). In this lengthy essay (for more concise versions, see Halffter 1965, 1968), he analyzed the differences between some general zoogeographic ideas based mostly on vertebrates with the patterns exhibited by insects, providing numerous examples of the latter. One example of such contrast is the Mexican Plateau, where vertebrate species are Nearctic but insect species are mostly Neotropical. Halffter (1964a) began his essay analyzing the origins of the American fauna. Based on Dunn (1931), he characterized three vertebrate "horofaunas" (this term following Smith 1949) that during the Cenozoic dispersed sequentially from Eurasia to North America and from there to South America: South American (more ancient), Northern Ancient (intermediate), and Holarctic (more recent). In each dispersal episode, taxa radiated into the new areas and induced the extinction of taxa belonging to the previous horofaunas. It is important to note that the term "dispersal" is used in its broadest sense, generally corresponding to the expansion of the distribution of a species not restricted by a barrier (Halffter and Morrone 2017). Instead of the north-to-south dispersal shown by vertebrates, Halffter postulated that the dispersal of the entomofauna of the Americas was south to north. This contrast is remarkable, because most of the South American insects dispersed into North America at the end of the Cenozoic, simultaneously with the dispersal of the Holarctic horofauna to Mexico, Central America, and northern South America.

When analyzing the phylogenetic affinities of the entomofauna, Halffter (1964a) postulated that the closest relatives of Neotropical insects were found in the southern continents, showing a Gondwanic distributional pattern. Halffter explicitly did not take part of the controversy on continental drift, although he noted that some zoogeographers (e.g., Jeannel 1942) supported such theory. René Jeannel's works have been an important influence for many entomologists (see Roig-Juñent 2005), contrasting with vertebrate zoologists who usually followed William D. Matthew, George G. Simpson, and Philip Darlington, Jr., the most conspicuous representatives of the traditional dispersalist approach, known as "Holarcticism" (Reig 1968; Halffter 1974) or the "New York school of zoogeography" (Croizat 1958, 1964; Morrone 2009). The strong phylogenetic affinities shown by insect taxa from South America, India, Australia, and New Zealand, however, made Halffter admit that Gondwanic distributions constituted a real pattern.

Halffter (1964a) postulated that the first dispersal of South American insects to North America, through the Central American bridge, might have occurred between the Late Cretaceous and Early Eocene, simultaneously with the first invasion of

North American vertebrates to South America (the South American horofauna). The main evidence of this event is the entomofauna of the Mexican Plateau and the southern United States, constituted by species belonging to South American genera. Halffter considered that this massive colonization could not have occurred under the current ecological and physiographic conditions of the Mexican Plateau and the barrier constituted by the Transmexican Volcanic Belt. The latter began its development during the Pliocene, so the dispersal of the South American entomofauna might have occurred before the Miocene. Additionally, Halffter noted that the south-to-north dispersal of insects was similar to the dispersal of several plant taxa. In the Pliocene, North and South America reconnected, but the physiography of the Mexican Plateau was similar to the modern one, so the insects (as well as some vertebrates) that dispersed from South America, originating the Typical Neotropical elements, distributed in the tropical forests of the coastal lowlands of the Pacific and Gulf of Mexico.

The next part of Halffter's (1964a) contribution analyzed the Mexican Transition Zone. He considered that it is an area of great biogeographic interest, although it is relatively unstudied because of its complexity and the poor knowledge of its fauna and flora. The center of the Mexican Transition Zone lies in the Sierra Madre Occidental, Sierra Madre Oriental, and Transmexican Volcanic Belt, the high valleys of Oaxaca, and the highlands of Chiapas and Guatemala, also including the lowlands surrounding them (Fig. 3.4). The delimitation of the Nearctic and

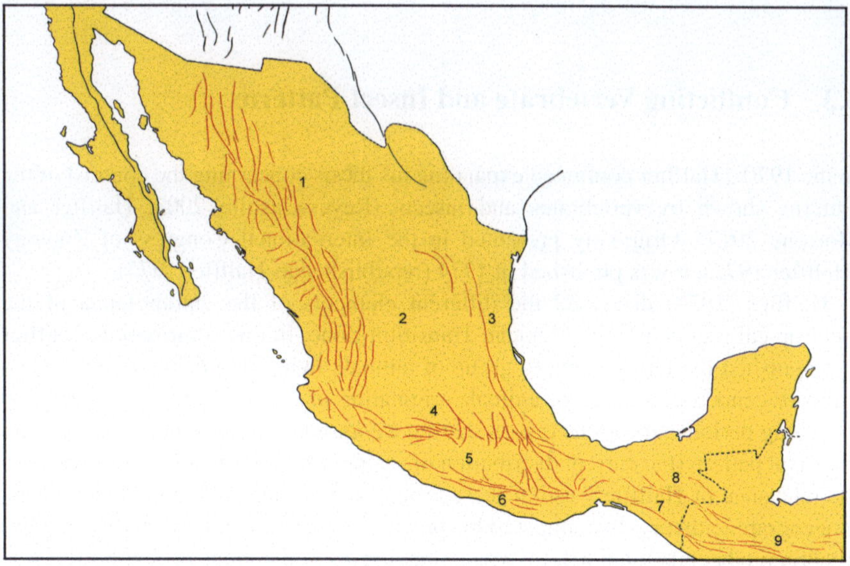

Fig. 3.4 Mexican Transition Zone (modified from Halffter 1964a). (1) Sierra Madre Occidental; (2) Sierra Madre Oriental; (3) Mexican Plateau; (4) Transmexican Volcanic Belt; (5) Balsas Basin; (6) Sierra Madre del Sur; (7) Chiapas Highlands; (8) Chiapas Central Massif; (9) Central American Nucleus

Neotropical zoogeographic regions is mostly based on vertebrates, but insects (and possibly other ancient groups) from Mexico and southern United States are different from the vertebrates characterizing them, because of different paleoecological and paleogeographical processes having affected them in the past. After reviewing previous zoogeographic regionalizations of Mexico based on vertebrates—especially those of Smith (1941), Goldman and Moore (1945), and Ryan (1963)—Halffter (1964a) referred to the dispersal patterns of the entomofauna of the Mexican Transition Zone. He confirmed the three patterns described by him in 1962 and added a new one for those septentrional elements that dispersed very early in the Cenozoic from Eurasia.

This newly described Paleoamerican dispersal pattern corresponds to septentrional (Holarctic) taxa that arrived into North America from Eurasia in the Early Cenozoic. Paleoamerican genera have species in the Mexican Plateau, the Transmexican Volcanic Belt and the Sierras Madre, and they also have species in the tropical lowlands. Most of them exhibit a notable and ancient dispersal to South America. Some of these genera are almost cosmopolitan, being more common in the tropics and temperate areas. Examples of Paleoamerican Scarabaeid genera provided by Halffter (1964a) are *Onthophagus* and *Copris*. Another example provided by Halffter was *Bombus* (Hymenoptera: Apidae), widely distributed in the Northern Hemisphere and with six species in Brazil. [Abrahamovich et al. (2004) cited 42 Neotropical and Andean species of this genus; it would be interesting to analyze whether the Andean species derived from the Neotropical taxa or resulted from a dispersal event from North America.]

3.3 Conflicting Vertebrate and Insect Patterns

In the 1970s, Halffter continued expanding his thesis concerning the contrast of the patterns shown by vertebrates and insects (Reyes-Castillo 2003; Halffter and Morrone 2017). Originally presented in the International Congress of Zoology (Halffter 1972), it was published in 1974 (republished as Halffter 1975).

Halffter (1974) discussed the different elements of the entomofauna of the Neotropical region and the Mexican Transition Zone. In this contribution, Halffter distinguished the terms fauna (a group of animals living in a defined area), horofauna or cenocron (a group of animals originating in a defined area and coexisting for a long period and for this reason sharing a common biogeographic history), and dispersal pattern (the current distribution of a cenocron). Zunino (2007) highlighted the relevance of Halffter's use of these terms, as a fundamental component of his biogeographic theory. In contrast to his previous contribution, here Halffter explicitly accepted continental drift to explain the origins of the fauna of South America.

Halffter (1974) characterized briefly the two alternative hypotheses that have been formulated to explain the distributional patterns of the South American fauna. Some authors ("courant holarcticiste") have postulated that all the South American taxa originated in the Old World and then dispersed to North America and finally arrived to South America. Other authors have postulated an origin in the other Austral continents. Instead of choosing one of them, Halffter enunciated a synthetic theory for the Mexican Transition Zone, accepting the coexistence of both Gondwanic taxa and Holarctic immigrants. Halffter (1974) considered that his synthetic ideas were similar to those of Rapoport (1968), Reig (1968) and Udvardy (1969).

Halffter (1974) classified the taxa distributed in the Mexican Transition Zone into four dispersal patterns:

Nearctic dispersal pattern: it includes recent Holarctic and some Nearctic taxa. In the Mexican Transition Zone, these taxa are generally restricted to areas above 1500 m.

Paleoamerican dispersal pattern: taxa distributed in lowlands and mountains, being the latter centers of diversification rather than dispersal routes. Within them, Halffter characterized two types of distributional patterns: relict species belonging to groups widely ranged in the tropics of the Old World and species belonging to groups widely distributed in North America. In the Mexican Transition Zone and South America, Paleoamerican taxa dispersed before the formation of the Mexican Plateau and the expansion of the deserts of western North America.

Mexican Plateau dispersal pattern and ancient South American (Neotropical) origin: taxa widely distributed in the Mexican Plateau and the highlands of Guatemala and Chiapas. These taxa are rarely found in the mountains of the Mexican Transition Zone, where Nearctic and Paleoamerican taxa are the most abundant.

Neotropical dispersal pattern and recent origin: groups that are distributed in the lowlands, which did not invade the Mexican highlands, and probably extended their ranges northward after the Pliocene.

Halffter (1976) discussed the four dispersal patterns of the Mexican Transition Zone, especially dealing with the relationships with the North American fauna. In addition to the dispersal patterns already recognized, Halffter considered that there were two patterns peripherically related to those in the transition zone: Greater Antilles relics and Old South American taxa from the United States. Within the Paleoamerican pattern, Halffter (1976) described two distributional varieties. One corresponds to relictual species from very localized areas, although their representatives in the Old World are widely distributed. The other corresponds to genera that have had important radiations, having representatives in the United States, Central America, South America (Andes from Colombia to Peru), and the Antilles. One example of a Paleoamerican taxon is *Copris* (Fig. 3.5). Within the Nearctic pattern, Halffter (1976) distinguished taxa with Nearctic affinities and those belonging to Holarctic lineages.

In 1978, Halffter described a fifth dispersal pattern for the Mexican Transition Zone. In addition to the previously recognized patterns, he considered it necessary to describe the Mountain Mesoamerican pattern. This pattern corresponds to elements of the mountain forests of the Central American Nucleus, which comprises the highlands of Chiapas, Guatemala, Honduras, El Salvador, and Nicaragua north

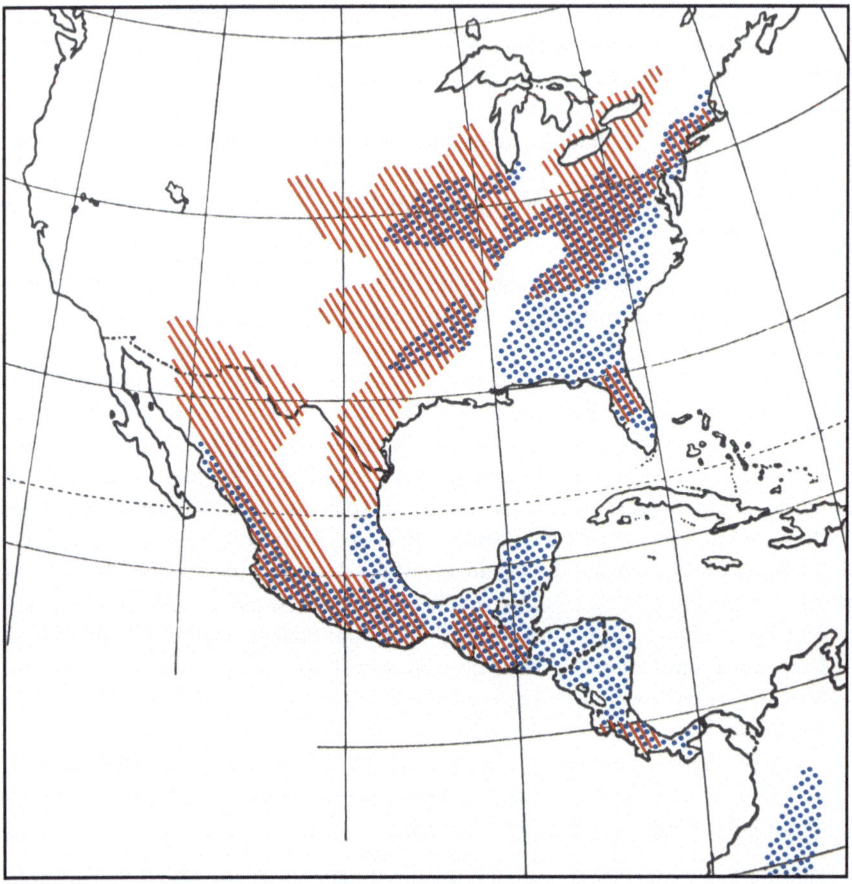

Fig. 3.5 Distribution of the genus *Copris* (Coleoptera: Scarabaeidae), a Paleoamerican taxon (modified from Halffter 1976). Blue circles, *C. minutus* species group; orange lines, *C. fricatus* species group

of Lake Nicaragua, and exists since the Late Cretaceous. Taxa assigned to the Mountain Mesoamerican pattern evolved in the Central American Nucleus and then dispersed northwest and southeast of it. Their affinities are ancient South American, and they are distributed mainly in mountain and cloud forests, penetrating occasionally in pine-oak forests. Halffter (1978) noted that insects following this pattern are quite scarce, compared to the other dispersal patterns of the Mexican Transition Zone, but several reptiles and amphibians showing this pattern are common.

Halffter (1978) characterized the history of the Mountain Mesoamerican pattern. During most of the Cretaceous, the southernmost part of North America (Oaxaca) was separated by 3000 km of ocean from northern South America. In the Late Cretaceous, there were some continental islands in the area of the Central American Nucleus, as well as some volcanic islands between Nicaragua and Panama. At the end of the Cretaceous, the Central American Nucleus raised definitely, as well as the northern Andes in South America. From this moment to the Miocene, dispersal between North and South America was only possible for taxa capable of "jumping islands." From the end of the Miocene to the Pliocene, southern Central America raised, and both continents united. During this period, South American taxa arrived into the Central American Nucleus and evolved in situ. From the Pliocene to recent, the elements evolved in the Central American Nucleus dispersed northward, into the mountains of the Mexican Transition Zone.

Halffter (1978) provided examples of vicariance of genera of Passalidae (Coleoptera): north-south (Central American Nucleus-mountains north of the Isthmus of Tehuantepec) in *Spurius*, *Oileus*, *Pseudacanthus*, and *Vindex* and east-west (both in the Central American Nucleus and in the mountains north of the Isthmus of Tehuantepec) in *Proculus*, *Chondrocephalus*, *Vindex*, *Undulifer*, *Proculejus*, and *Petrejoides*. The distributional patterns of the tribe Proculini (Coleoptera: Passalidae) were analyzed by Reyes-Castillo and Halffter (1978).

3.4 The Mountain Entomofauna

In the late 1980s, Halffter focused his attention on the mountain entomofauna (Morrone 2015a). The paper published in 1987 (translated to Spanish: Halffter 2003, 2006) can be considered as inaugurating this period. Halffter (1987) provided a general overview of his previous ideas on the mountain entomofauna of the Mexican Transition Zone. Halffter clarified his use of the term "pattern" or "distributional pattern":

> In my work I frequently use the concept of pattern…, which corresponds to a synthesis of the essential features of the distribution of a set of coexisting organisms that originated or became integrated in a given area and time, are subjected to the same macroecological pressures for a prolonged period, live under the same physiographic conditions, and have a common biogeographic history. The concept of pattern is a generalization. It is intended for reference to help us analyze and compare differences in the distribution of each taxon. (Halffter 1987, pp. 96–97)

After a brief description of the major mountain systems of the Mexican Transition Zone, Halffter (1987) presented his theory. He postulated that the montane insects of the Mexican Transition Zone belong to two groups: one that occupies the ranges north of the Isthmus of Tehuantepec and the other that occupies the mountain systems south of the isthmus (Sierra Madre de Chiapas and mountains of Central America extending to the Nicaraguan depression). Most of the montane insects north of the Isthmus of Tehuantepec are of northern (Nearctic) origin, and the mountain ranges have allowed their dispersal southward. These mountain ranges, especially the Transmexican Volcanic Belt and the mountains of Oaxaca and Guerrero, have been frequently areas of isolation and vicariance. The northern elements fit into the Nearctic or Paleoamerican distributional patterns, according to the time of their arrival to the Mexican Transition Zone. The mountains north of the

Isthmus of Tehuantepec also include Mesoamerican elements, whereas insects from the tropical lowlands and from the Mexican Plateau are rare. The mountains north of the isthmus have been, in general, a poor route for the dispersal of the more modern Nearctic fauna, particularly the hygrophilous elements whose expansion is hampered by the lack of lakes and rivers in the Mexican ranges, particularly in the Sierra Madre Occidental. Most of the insects of the mountains south of the Isthmus evolved in the Central American Nucleus, which received an ancient and very important contribution from South America and also acquired some northern elements. The fauna that evolved in the Nucleus expanded toward South America and toward the north through the Isthmus of Tehuantepec, corresponding to the Mesoamerican distributional pattern. South of the Mexican Transition Zone, in the Sierra de Talamanca of Costa Rica and Panama, mountains were colonized largely by elements from the tropical lowlands.

In the following two decades, Halffter's interest focused on the altitudinal patterns shown by the mountain entomofauna of the Mexican Transition Zone (Halffter et al. 1995, 2008b; Lobo and Halffter 2000; Escobar et al. 2007). Halffter et al. (1995) analyzed the distributional patterns of three groups of coprophagous and necrophagous beetles (Scarabaeinae, Geotrupinae, and Silphidae) along an altitudinal transect in the state of Veracruz, from sea level to 4282 m altitude. They found that the Paleoamerican, Nearctic, Mexican Plateau, Neotropical, and Mountain Mesoamerican patterns corresponded to well-defined altitudinal zones, with an overlap of elements with Neotropical and Nearctic affinities, resulting in a transitional area (a similar pattern was detected for bird taxa; Sánchez-González and Navarro-Sigüenza 2009). Taxa assigned to the Paleoamerican distributional pattern, however, did not exhibit common ecological features. Their diversification was due to their ancient history in the Mexican Transition Zone and their adaptive plasticity (for an exhaustive biogeographic analysis of a taxon showing the heterogeneity of the Paleoamerican pattern, see Zunino and Halffter 1988). Thus, Halffter et al. (1995) considered it better to split the Paleoamerican pattern into four different distributional subpatterns:

Relict Paleoamerican subpattern: species of genera with wide geographical and ecological distribution in the Old World but represented in the Mexican Transition Zone by endemic species with very restricted distributions.

Mountain Paleoamerican subpattern: lineages that have colonized successfully the mountains of the Mexican Transition Zone and, to a lesser extent, those of Central America. They have undergone significant speciation in the mountains, driven by vicariance.

Mountain Mesoamerican subpattern (Paleoamerican elements): species following the Mesoamerican Mountain pattern but being of Paleoamerican affinity (Old World).

Tropical Paleoamerican subpattern: species found in tropical lowlands and moderate altitudes. Their distribution may be very similar to those of the Neotropical pattern, but their affinities are Paleoamerican (Old World).

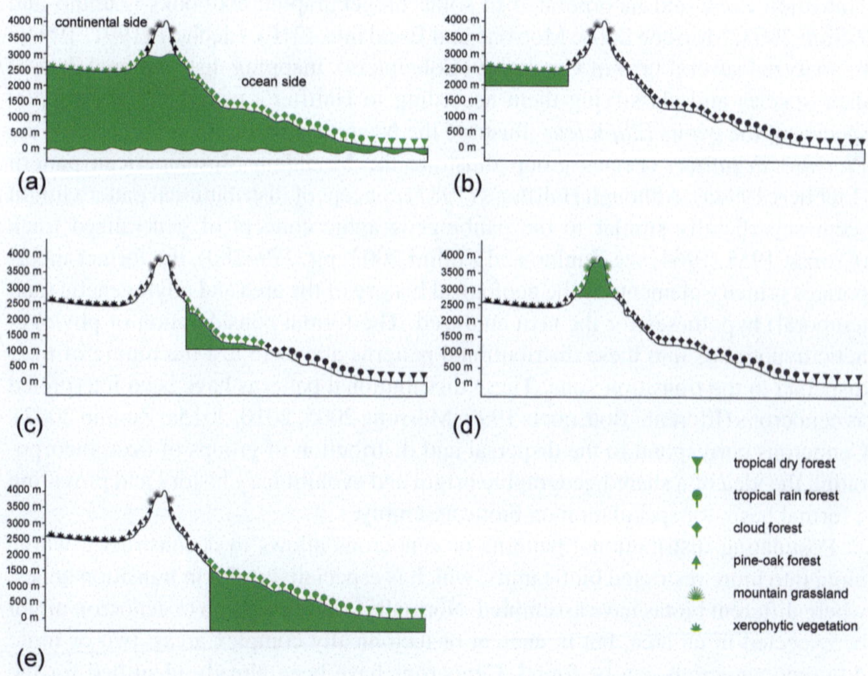

Fig. 3.6 Distributional patterns and vegetation types that correspond to the different cenocrons, showing the altitude and continental and external slopes. (**a**) Paleoamerican biota; (**b**) Mexican Plateau cenocron; (**c**) Mountain Mesoamerican cenocron; (**d**) Nearctic cenocron; (**e**) Typical Neotropical cenocron

Halffter et al. (1995) found that the altitudinal pattern parallels the latitudinal one. At lower altitudes, the tropical (Neotropical and Tropical Paleoamerican) elements are more abundant, whereas at higher altitudes, Nearctic taxa predominate. These elements are associated to specific vegetation types (Fig. 3.6).

3.5 Halffter's Distributional Patterns and Biotic Assembly

Halffter (1972, 1974, 1976) used the term "dispersal pattern," which he defined as follows:

> By dispersal pattern (Halffter 1972, 1974) I mean the present distribution of a cenocron. The term cenocron, coined by Reig (1962, 1968) is equivalent to the better-known term horofauna, and refers to a group of organisms which originated in or become integrated as such in a given area, which have coexisted for a prolonged period and have a common biogeographic history. (Halffter 1976, p. 5)

Halffter (1987) preferred to use the term "distributional pattern" for the same concept. It has been widely used by several authors working on the Mexican

Transition Zone and incorporated to some biogeographic textbooks (Zunino and Zullini 2003; Morrone 2009; Morrone and Escalante 2016). Liebherr (1991, 1994a, b) analyzed several taxa of Carabidae (Coleoptera), mapping distributional data of their species and classifying them according to Halffter's patterns. For example, species of the genus *Elliptoleus* illustrate the Nearctic pattern, whereas those of the *Platynus degallieri* species group illustrate the Mountain Mesoamerican pattern (Liebherr 1994a). Although Halffter's (1987) concept of distributional pattern might seem superficially similar to the panbiogeographic concept of generalized track (Croizat 1958, 1964; see Zunino and Zullini 2003, pp. 279–280), the former incorporates primary elements of the geological history of the area and phylogenetic (and temporal) hypotheses for the taxa analyzed. The formal consideration of phylogenetic hypotheses into these distributional patterns allows to test the timing of their dispersal to the transition zone. These distributional patterns have been interpreted as cenocrons (Llorente Bousquets 1996; Morrone 2005, 2010, 2015a; Zunino 2007). Cenocrons correspond to the dispersal and distribution of groups of taxa, incorporating the idea of a shared geographic origin and evolutionary history and providing a formal basis for speculation on biotic assembly.

Postulating distributional patterns or cenocrons allows to deconstruct a whole biota into more restricted biotic units, which is especially useful in transition zones, where different biotas have assembled. Normally, a single pattern or cenocron might be expected in an area, but in ancient or tectonically complex areas, two or more different cenocrons can be found. Cenocrons have been already identified for the Mexican Transition Zone (Morrone 2015a), and similar units have been identified in the South American Transition Zone (Roig-Juñent et al. 2008, 2018). Lobo (2007) applied the concept of distributional pattern to the Iberian Scarabaeinae:

> After revising the main suggestions about the role played by phylogenetic inertia and the ecogeographical adaptations of species in connection with the place of origin of the lineage to which they belong, the 'dispersion pattern' concept proposed by Gonzalo Halffter is commented upon, highlighting the common point of view of these approaches, i.e. the use of present-day biogeographic and ecological data to discern the historical processes underlying current biodiversity patterns. Subsequently, taxonomic, ecological and biogeographical information on Scarabaeine dung beetles is used to discriminate groups of species that probably share a common history of colonization for the Iberian Peninsula. Lastly, both the biogeographic history of these dung beetles and the history of land masses are revised to propose a coherent narrative explanation for the origin of the detected dispersion patterns. (Lobo 2007, p. 159)

For the Iberian Peninsula, which has been assigned either to the Euro-Mediterranean Transition Zone (Zunino 1985) or to the Palearctic region (Morrone 2015c), Lobo (2007) identified five dispersal patterns. I fully agree with this author, when he states that:

> The current distribution of any species is the spatial result of the interaction between the ecophysiological adaptations that it possess, the environmental and topographical characteristics of the territory that it inhabits, and the unique and unrepeatable histories of both the taxon and the earth. Understanding the genesis of this distribution requires knowing how these three sources of variability interact and have interacted in the past. (Lobo 2007, p. 160)

Moctezuma et al. (2016) and Halffter and Morrone (2017) hypothesized that the taxa belonging to the same cenocron tend to show "ecological inertia," being restricted to habitats that are similar to those of the area where they originally evolved. For example, in the highest parts of the mountains of the Mexican Transition Zone, Scarabaeid taxa that correspond to the Nearctic cenocron (and secondarily Mountain Paleoamerican taxa) respond positively to the reduction of the vegetation cover (Moctezuma et al. 2016). This is related to the phylogenetic origin of these taxa that belong to Holarctic or Nearctic groups where forest-adapted species are relatively scarce and most of the species are adapted to primary or secondary heliophylic habitats. On the contrary, species that belong to the Typical Neotropical cenocron inhabit mostly tropical forests, because they derive from forest taxa that evolved in northern South America and dispersed to the Mexican Transition Zone through Central America.

3.6 The Mexican Transition Zone in the Twenty-First Century

The Mexican Transition Zone as defined by Halffter includes Mexico, southern United States, and northern Central America. As the biota inhabiting northern Mexico and southern United States is clearly Nearctic, and the biota of the lowlands of southern Mexico and Central America are clearly Neotropical, I (Morrone 2005, 2006, 2014a, b) prefer to restrict the Mexican Transition Zone to the highlands of Mexico, Guatemala, Honduras, El Salvador, and Nicaragua. This Mexican Transition Zone in the strict sense includes the biogeographic provinces of the Sierra Madre Occidental, Sierra Madre Oriental, Transmexican Volcanic Belt, Sierra Madre del Sur, and Chiapas Highlands (see Chap. 4). From the perspective of biogeographic regionalization, assigning these provinces to the Mexican Transition Zone implies that they do belong simultaneously to the Nearctic and Neotropical regions (Morrone 2014b, 2017).

In a historical review of the Mexican Transition Zone, I (Morrone 2015a) considered that the distributional patterns recognized by Halffter and collaborators represented cenocrons. If the concept of cenocron is applied strictly to a set of taxa that dispersed into a biota (Morrone 2014c), only the Mexican Plateau, Mountain Mesoamerican, Nearctic, and Typical Neotropical qualify as cenocrons, whereas the Paleoamerican pattern represents the original biota of the Mexican Transition Zone. A summary of the development of the Mexican Transition Zone (Morrone 2015a) consists of the following stages (Fig. 3.7):

Jurassic-Cretaceous: the Paleoamerican biota extends in Mexico. It represents the original (Holarctic) biota of the country, and its taxa have evolved for a long period of time and have been affected by the vicariance and biogeographic convergence events that shaped the complex biotic history of the country.

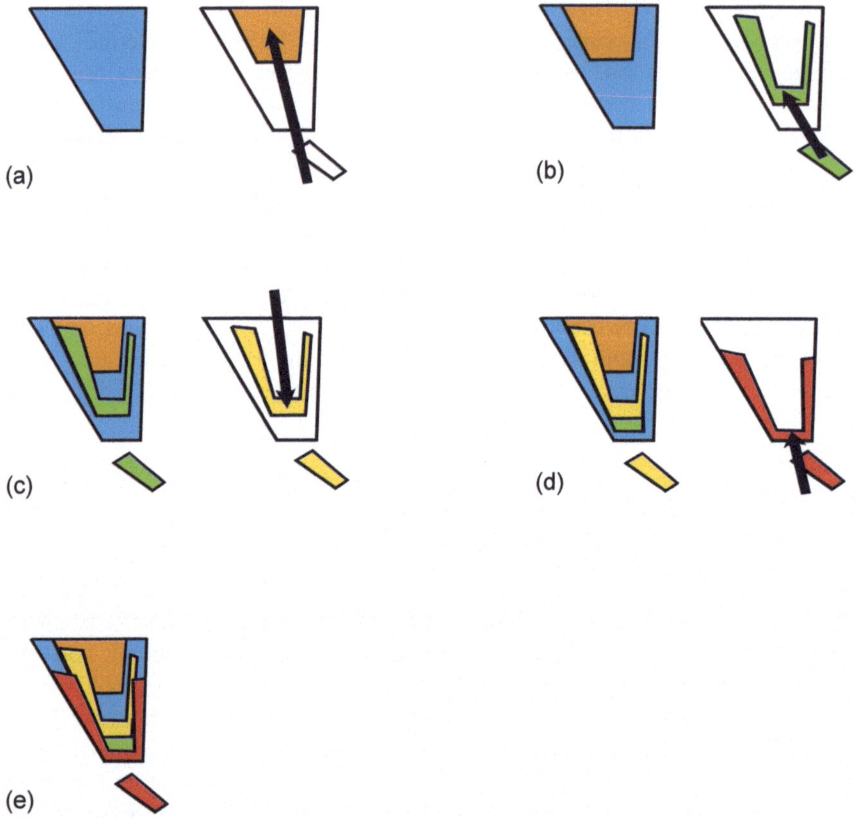

Fig. 3.7 Diagrammatic representation of the development of the Mexican Transition Zone (after Morrone 2015a). (**a**) Jurassic-Cretaceous, Paleoamerican biota (blue), and Late Cretaceous-Paleocene, dispersal of the Mexican Plateau cenocron (orange); (**b**) Oligocene-Miocene, dispersal of the Mountain Mesoamerican cenocron (green); (**c**) Miocene-Pliocene, dispersal of the Nearctic cenocron (yellow); (**d**) Pliocene-Pleistocene, dispersal of the Typical Neotropical cenocron (red); (**e**) recent, overlap of the different cenocrons

Late Cretaceous-Paleocene: dispersal from South America of the Mexican Plateau cenocron. It includes taxa widely distributed in the Mexican Plateau, which are rare in the mountains of the Mexican Transition Zone.

Oligocene-Miocene: dispersal from South America to the Central American Nucleus of the Mountain Mesoamerican cenocron. These taxa, with ancient South American affinities, evolved in the highlands of Chiapas, Guatemala, Honduras, El Salvador, and Nicaragua north of Lake Nicaragua and then dispersed northward to Mexico in the Pliocene.

Miocene-Pliocene: dispersal from North America to the Mexican Transition Zone of the Nearctic cenocron, which includes Holarctic and Nearctic taxa. These taxa are generally restricted to areas above 1500 m, in temperate and cloud forests.

Pliocene-Pleistocene: dispersal from South America to the Mexican Transition Zone of the Typical Neotropical cenocron. It corresponds to taxa distributed in tropical forests of the lowlands that were not able to invade the highlands.

Halffter and Morrone (2017) reviewed the distributional patterns and cenocrons of the Mexican Transition Zone and discussed their heuristic value. They analyzed three case studies in some detail: the Neotropical genus *Canthon* and the tribe Phanaeini, and the Holarctic subfamily Geotrupinae.

Halffter et al. (2019) analyzed the *Onthophagus chevrolati* species group (Coleoptera: Scarabaeidae) as a case study of a Paleoamerican taxon diversifying in the Mexican Transition Zone. They analyzed the dispersal and vicariance events occurring specially in the Transmexican Volcanic Belt, which they considered has a central role in the biotic assembly of the Mexican Transition Zone. Additionally, they renamed the "Mountain Mesoamerican subpattern (Paleoamerican elements)" recognized by Halffter et al. (1995) to "Mesoamerican Paleoamerican subpattern."

3.7 Impact of Halffter's Theory

Halffter's ideas concerning the Mexican Transition Zone have been considered to represent the classical dispersalist approach, but this is an oversimplification (Morrone and Márquez 2001; Morrone 2015a). Although in his contributions of the 1960s Halffter postulated dispersal patterns, his mature theory (e.g., Halffter 1987, 2017; Halffter et al. 1995, 2019) considers both processes of dispersal and vicariance. It is interesting that both Reig (1962, 1981) and Savage (1966, 1982) also passed through a similar evolution in their biogeographical conceptions (Morrone 2003). In contrast to other authors that in the 1980s crusaded for either dispersal or vicariance, Halffter, Reig, and Savage chose a "middle way," anticipating the dispersal-vicariance model promoted during the last two decades (Brooks 2004; Lieberman 2004; Sanmartín and Ronquist 2004; Riddle et al. 2008; Crisci and Katinas 2009; Morrone 2009, 2011). Instead of assuming that dispersal and vicariance are the only drivers of biotic assembly, as do dispersalists, panbiogeographers, and cladistic biogeographers, the dispersal-vicariance model assumes that both processes are relevant and should be analyzed. Dispersal occurs normally and is a prerequisite for vicariance. Additionally, once biotas have been identified and represented as generalized tracks or areas of endemism, the identification of cenocrons may allow determining the timing of their dispersal (Morrone 2009, 2015b).

Since its original formulation half a century ago, Halffter's transition zone has given major impetus to the study of biogeographic patterns in Mexico (Reyes-Castillo 2003; Zunino 2007; Morrone 2015a). Different authors have studied the Mexican Transition Zone, using the most varied approaches and methods.

Dispersal Analyses Most of the earliest authors that applied Halffter's ideas to the distributional analyses of plant and animal taxa worked under a dispersalist framework (e.g., Evans 1966; Matthews 1966; Ball 1968; Rapoport 1968; Udvardy 1969;

Reyes-Castillo 1970; Martins 1971; Mateu 1974; Reichardt 1977; Axelrod 1979; MacVean and Schuster 1981; Castillo and Reyes-Castillo 1984). In addition, other analyses placed in a dispersal-vicariance framework have used cladograms to analyze the dispersal of particular taxa in the Mexican Transition Zone (e.g., Kohlmann and Halffter 1988, 1990; Lanteri 1990; Daza et al. 2009; Rosas et al. 2011a; Zamora-Tavares et al. 2016).

Track Analyses Several authors have undertaken track analyses of the Mexican Transition Zone (Morrone and Márquez 2001; Escalante et al. 2004, 2018; Morrone and Gutiérrez 2005; Andrés Hernández et al. 2006; Huidobro et al. 2006; Toledo et al. 2007; Espinosa Organista et al. 2008; García-Marmolejo et al. 2008; Corona et al. 2009; Rosas et al. 2011b; García Díaz et al. 2015; González-Ávila et al. 2017). These authors have analyzed different taxa (plants, insects, crustaceans, fish, and mammals), finding that some of the generalized tracks identified correspond to Halffter's patterns, but in other cases new patterns have emerged. Morrone and Márquez's (2001) track analysis of 134 beetle (Coleoptera) taxa from the Mexican Transition Zone and other areas identified two generalized tracks. The northern generalized track comprises taxa distributed basically in mountain areas within the Sierra Madre Occidental, the Sierra Madre Oriental, the Transmexican Volcanic Belt, the Balsas Basin, and the Sierra Madre del Sur. The southern generalized track comprises taxa distributed in the Sierra Madre de Chiapas and lowland areas in Chiapas, the Mexican Gulf, and the Mexican Pacific Coast, reaching in the south the Panamanian Isthmus.

Escalante et al. (2004) analyzed the distributional patterns of 46 Mexican Nearctic mammal species and identified generalized tracks and nodes, in order to determine the southernmost boundary of the Nearctic region in Mexico. They found six generalized tracks and nine nodes, the latter located basically in the Sierra Madre Oriental, Transmexican Volcanic Belt, Sierra Madre del Sur, and Chiapas biogeographic provinces. The highlands of Chiapas were found to represent the southernmost area inhabited by Nearctic taxa. The other biogeographical provinces, together with the Sierra Madre Occidental and the Balsas Basin provinces, represent the Mexican Transition Zone in the strict sense. They concluded that the Mexican Transition Zone represents an evolutionarily "active" zone, where several speciation events have taken place in the past.

Espinosa Organista et al. (2008) undertook a detailed analysis based on plants, insects, and vertebrates, finding six patterns that are confined to specific mountain slopes (Fig. 3.8): (1) coastal-mountain pattern, in the coastal slopes of the Sierra Madre Occidental, Sierra Madre Oriental, Sierra Madre del Sur, Chiapas Highlands, and western and eastern extremes of the Transmexican Volcanic Belt; (2) circum-Balsas river basin sub-humid mountain pattern, in the Transmexican Volcanic Belt and Sierra Madre del Sur, predominantly in the slopes oriented to the Balsas river basin; (3) circum-Plateau semiarid and arid mountain pattern, in the mountains that surround the Mexican Plateau; (4) mountain cloud pattern, distributed discontinuously in the most humid areas of the Gulf of Mexico slope; (5) Gulf slope mountain

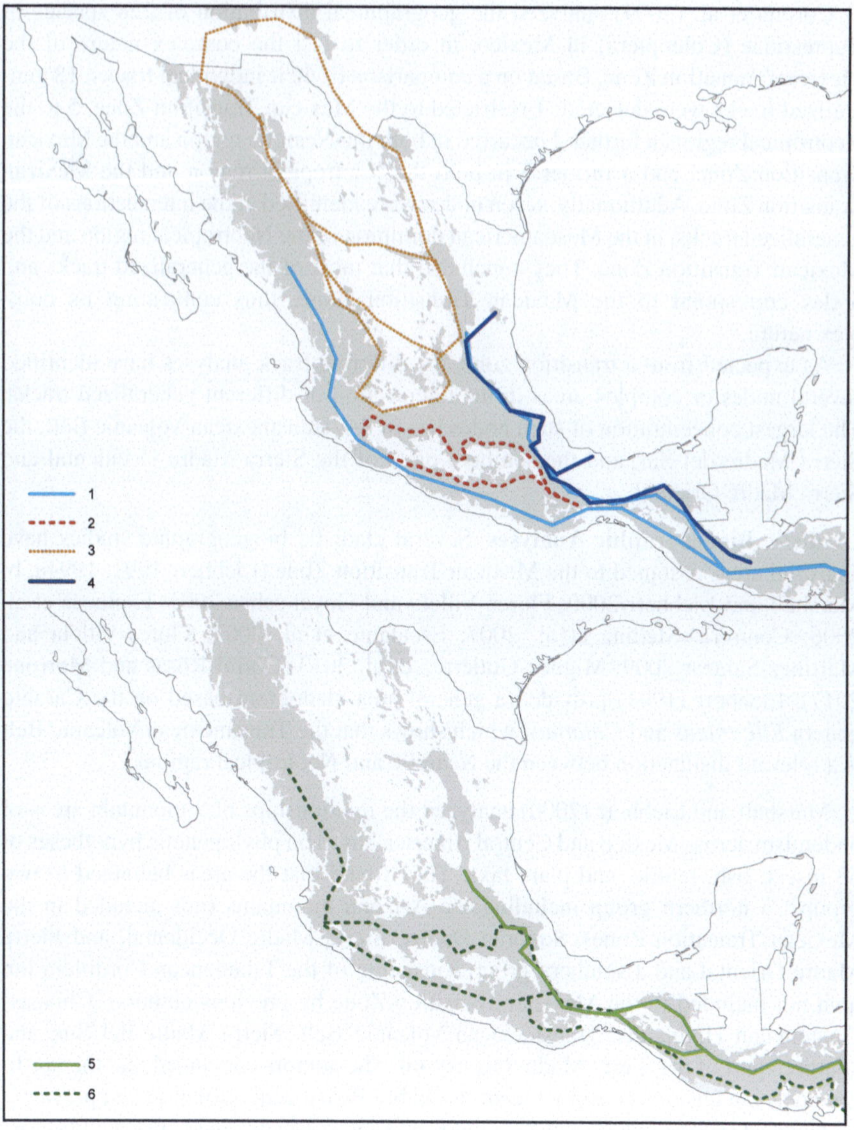

Fig. 3.8 Main generalized tracks obtained by Espinosa Organista et al. (2008). (1) coastal-mountain; (2) circum-Balsas river basin sub-humid mountain; (3) circum-Plateau semiarid and arid mountain; (4) mountain cloud; (5) Gulf slope mountain; (6) Pacific sub-humid mountain

pattern, in the Gulf of Mexico mountain slopes and in the transition between pine-oak forests, cloud forests, and rain forests; and (6) Pacific sub-humid mountain pattern, in the ecotone between pine-oak forests and tropical deciduous forests of the Pacific slope.

Corona et al. (2009) analyzed the geographical distribution of 228 species of Buprestidae (Coleoptera) in Mexico, in order to test the complex nature of the Mexican Transition Zone. Based on a comparison of their individual tracks, 13 generalized tracks were detected: 1 restricted to the Mexican Transition Zone, 5 to the Neotropical region, a further 2 occurred in both the Nearctic region and the Mexican Transition Zone, and a further 5 in both the Neotropical region and the Mexican Transition Zone. Additionally, seven nodes were identified at the intersections of the generalized tracks, in the Mesoamerican dominion of the Neotropical region and the Mexican Transition Zone. They concluded that most of the generalized tracks and nodes correspond to the Mexican Transition Zone, thus confirming its complex nature.

As expected from a transition zone, the different track analyses have identified several nodes or complex areas in the intersection of different generalized tracks. The largest concentration of such nodes lies in the Transmexican Volcanic Belt, the Sierra Madre del Sur, and the southern parts of the Sierra Madre Occidental and Sierra Madre Oriental.

Cladistic Biogeographic Analyses Several cladistic biogeographic studies have analyzed areas assigned to the Mexican Transition Zone (Liebherr 1991, 1994a, b; Marshall and Liebherr 2000; Flores-Villela and Goyenechea 2001; Espinosa et al. 2006; Contreras-Medina et al. 2007; Escalante et al. 2007; Flores-Villela and Martínez-Salazar 2009; Miguez-Gutiérrez et al. 2013; Corral-Rosas and Morrone 2017). Liebherr (1991) provided a general area cladogram based on the Carabid genera *Elliptoleus* and *Calathus*, which shows that the Transmexican Volcanic Belt is a relevant disjunction between the Nearctic and Neotropical regions.

Marshall and Liebherr (2000) analyzed the relationships of 9 mountain areas of endemism across Mexico and Central America based on phylogenetic hypotheses of 33 insect, fish, reptile, and plant taxa. They found that the areas belonged to two groups: a northern group including the Arizona mountains (not included in the Mexican Transition Zone), Sonoran Desert, Sierra Madre Occidental, and Sierra Madre Oriental and a southern group consisting of the Talamancan Cordillera (an area not included in the Mexican Transition Zone by previous authors), Chiapas-Guatemalan Highlands, Transmexican Volcanic Belt, Sierra Madre del Sur, and southern part of the Sierra Madre Occidental. The authors concluded that the northern areas are characterized by recent, probably Pleistocene, isolation and prevalent widespread species, whereas the southern areas probably diverged after the Pliocene closure of the Panamanian Isthmus.

Miguez-Gutiérrez et al. (2013) analyzed the relationships of the areas of endemism within the Mexican Transition Zone based on three different biogeographic regionalizations (Marshall and Liebherr 2000; Flores-Villela and Goyenechea 2001; Morrone 2006). They constructed taxon-area cladograms for ten genera of beetles, gymnosperms, lizards, and snakes, using the areas of the three regionalizations, and then obtained the general area cladograms (Fig. 3.9). They found two groups of

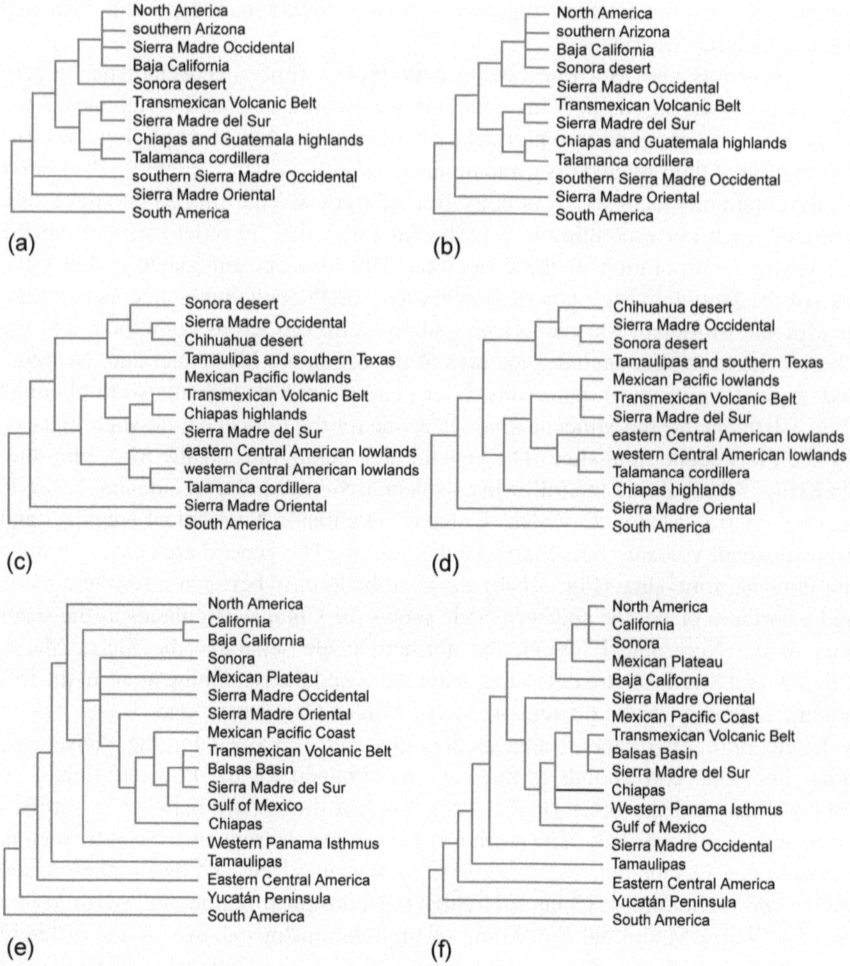

Fig. 3.9 General area cladograms evaluated by Miguez-Gutiérrez et al. (2013). (**a**) Marshall and Liebherr's (2000) areas, assumption 0; (**b**) Marshall and Liebherr's (2000) areas, assumption 1; (**c**) Flores-Villela and Goyenechea's (2001) areas, assumption 0; (**d**) Flores-Villela and Goyenechea's (2001) areas, assumption 1; (**e**) Morrone's (2006) areas, assumption 0; (**f**) Morrone's (2006) areas, assumption 1

areas, one with Neotropical affinities and the other with Nearctic affinities. Some common patterns were the close relationships between the Sierra Madre del Sur and the Transmexican Volcanic Belt and between the Chiapas Highlands and the Talamanca ridge. Miguez-Gutiérrez et al. (2013) concluded that the most important vicariance events within the Mexican Transition Zone were the elevation of the Transmexican Volcanic Belt, which divided the areas with Nearctic affinities in the north and those with Neotropical affinities in the south, and the rising of the

Tehuantepec Isthmus, which together with the Nicaraguan depression isolated Nuclear Central America.

Corral-Rosas and Morrone (2017) undertook a time-sliced cladistic biogeographic analysis, based on 49 taxa, including plants, insects, and vertebrates. Each of the taxa was assigned to a particular cenocron, based on their phylogenies, the distribution of their related taxa, and molecular dating (when available). The matrix for the parsimony analysis of paralogy-free subtrees was partitioned into three submatrices, each corresponding to a particular time-slice, in order to represent the successive incorporation of the cenocrons. The Miocene time-slice includes the taxa of the Mountain Mesoamerican cenocron, the Pliocene time-slice includes the taxa of the Mountain Mesoamerican and Nearctic cenocrons combined, and the Pleistocene time-slice includes the taxa of the Mountain Mesoamerican, Nearctic, and Typical Neotropical cenocrons. Four general area cladograms were obtained (Fig. 3.10): one for the Miocene time-slice, one for the Pliocene time-slice, and two for the Pleistocene time-slice. The general area cladogram for the Miocene time-slice (Fig. 3.10a) shows the following sequence: Sierra Madre Occidental, Nearctic region, Sierra Madre Oriental, Chiapas Highlands-Neotropical region, and Transmexican Volcanic Belt-Sierra Madre del Sur. The general area cladogram for the Pliocene time-slice (Fig. 3.10b) shows a dichotomy between a southern clade and a northern clade; the southern clade shows the Chiapas Highlands as the sister area of the Neotropical region; the northern clade separates the Sierra Madre Oriental and Sierra Madre del Sur as sister areas and the remaining areas in the following sequence: Nearctic region, Sierra Madre Occidental, and Transmexican Volcanic Belt. Two general area cladograms were obtained for the Pleistocene time-slice. One corresponds to the same as obtained for the Pliocene time-slice (Fig. 3.10b), and the other (Fig. 3.10c) shows a dichotomy between a northern clade and a southern clade. The northern clade shows the sequence Nearctic region, Sierra Madre Occidental, and Transmexican Volcanic Belt, and the southern clade shows two dichotomies: Chiapas Highlands-Neotropical region and Sierra Madre Oriental-Sierra Madre del Sur. Some of the relationships shown by the different general area cladograms are coincidental. The close relationship between the Chiapas Highlands and the Neotropical region is shown in the general area cladograms of the three time-slices; this situation might be because the Chiapas highlands are situated in the southernmost part of the Mexican Transition Zone, thus having a stronger Neotropical influence. A close relationship between the Sierra Madre Oriental and Sierra Madre del Sur is shown by the Pliocene and Pleistocene time-slices.

As expected from the complexity of a transition zone, results of the published cladistic biogeographic analyses differ widely, showing conflicting results when different areas of endemism and taxa are analyzed.

Endemicity Analyses Luna Vega et al. (1999), Escalante et al. (2005), Gutiérrez-Velázquez et al. (2013), Kobelkowsky-Vidrio et al. (2014), Munguía-Lino et al. (2016), García-Navarrete and Morrone (2018), Pinilla-Buitrago et al. (2018),

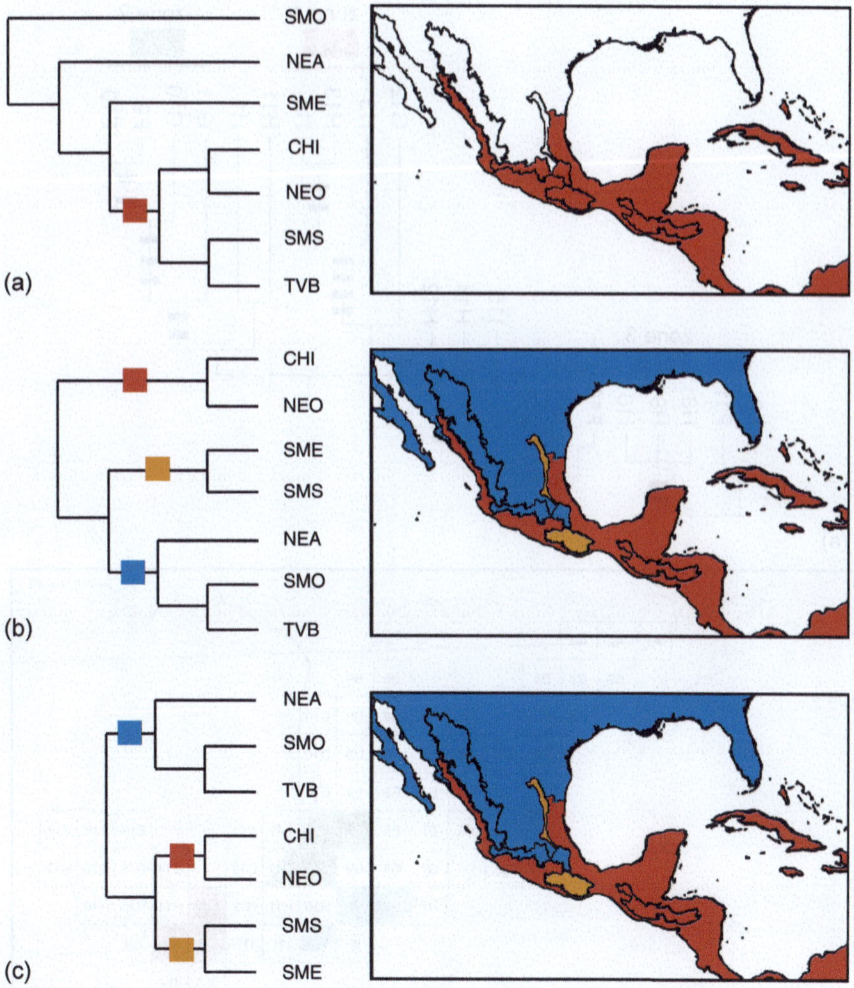

Fig. 3.10 General area cladograms obtained by Corral-Rosas and Morrone (2017) and the maps representing geographically the main clades. (**a**) Miocene time-slice; (**b**) Pliocene and Pleistocene time-slices; (**c**) Pleistocene time-slice

Rodríguez et al. (2018), Estrada Sánchez et al. (2019), and Santiago-Alvarado (2019) undertook analyses of endemicity. Gutiérrez-Velázquez et al. (2013) re-evaluated the biogeographical patterns in the Mexican Transition Zone through the recognition of congruence in the geographic distributions of Passalidae. They undertook a parsimony analysis of endemicity of specific distribution data, applying a null model of significant co-occurrence to the distributional data and using predicted potential distributions through ecological niche modeling. They found three

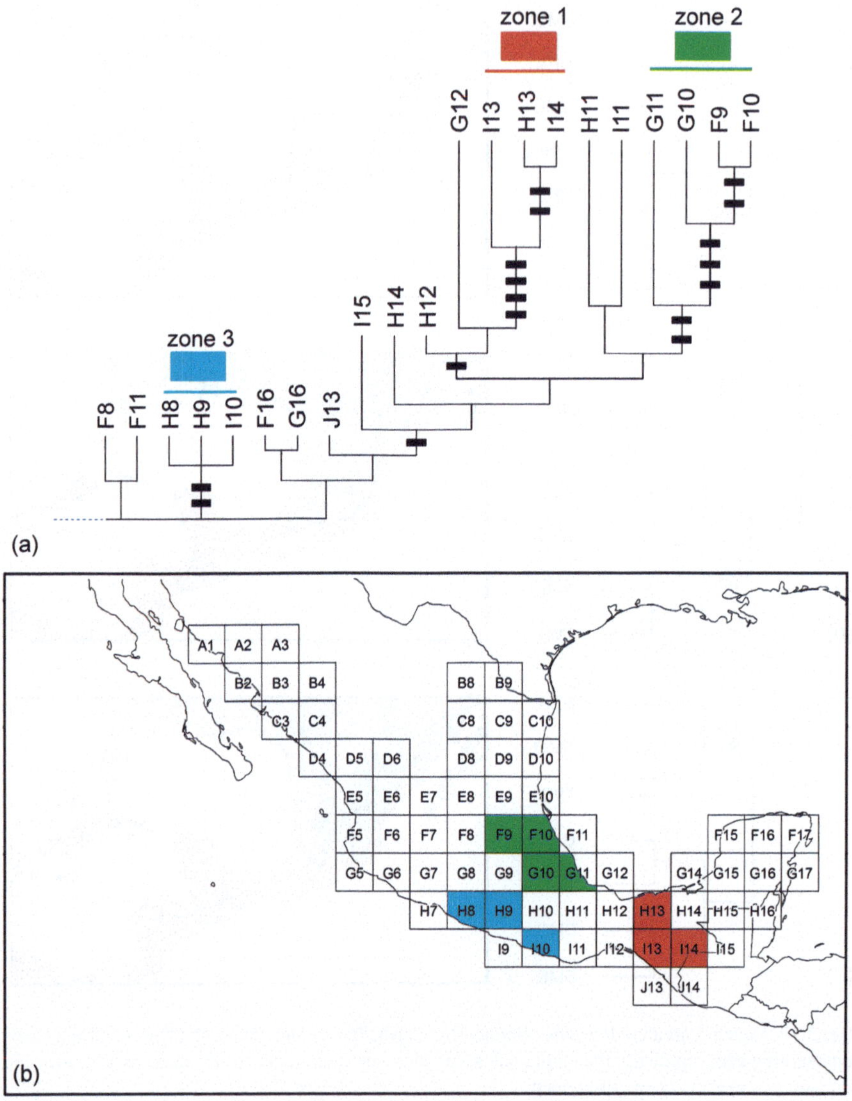

Fig. 3.11 Parsimony analysis of endemicity of Coleoptera undertook by Gutiérrez-Velázquez et al. (2013). (**a**) Area cladogram obtained; (**b**) map with the three zones represented

"general zones": general zone 1 includes species of restricted distribution in the state of Chiapas; general zone 2 includes species of restricted distribution located mainly in the states of Veracruz, Puebla, Hidalgo, and Querétaro; and general zone 3 includes species with restricted distribution occurring mainly in the states of Guerrero and Oaxaca (Fig. 3.11). These patterns resulted highly coincident with

Halffter's patterns: general zone 1 has species endemic to humid tropical forests of low, medium, and high altitude (800–3000 m) of Nuclear Central America (from Chiapas to Nicaragua), corresponding to Halffter's Mountain Mesoamerican and Typical Neotropical patterns, whereas the other zones correspond to the Mountain Mesoamerican pattern.

Phylogeographic Analyses There are many intraspecific phylogeographic analyses of taxa distributed in the Mexican Transition Zone (e.g., Navarro-Sigüenza et al. 2008; Vázquez-Miranda et al. 2009; Daza et al. 2010; Bryson Jr et al. 2011; Cox et al. 2012; Agorreta et al. 2013; Almendra et al. 2014; Rodríguez-Gómez and Ornelas 2015; Nolasco-Soto et al. 2017; Gutiérrez-Ortega et al. 2018). Bryson Jr et al. (2011) analyzed the phylogeographic structure and estimates of divergence times of three species of *Pituophis* (Colubridae) from the Mexican Transition Zone, to evaluate the postulated association of three Neogene geological events (the marine seaway inundation of the Isthmus of Tehuantepec, the formation of the Transmexican Volcanic Belt, and the secondary uplifting of the Sierra Madre Occidental) and of Pleistocene climate changes with their diversification. They identified three major lineages, with a strong phylogeographical structure within them. The estimated divergence dates and the geographical distribution supported mixed responses to these geological events. The phylogeographical structure highlights the influence of both Neogene vicariance events and Pleistocene climate change in shaping genetic diversity in this region. Despite the presence of two major geographical barriers in southern Mexico, extreme geological and environmental heterogeneity in this area may have differentially structured genetic diversity in highland taxa. To the north, co-distributed taxa may display a more predictable pattern of diversification across the warm desert regions.

Rodríguez-Gómez and Ornelas (2015) analyzed the role of past geological and climatic changes in the speciation of the hummingbird species *Amazilia violiceps*, *Amazilia viridifrons*, and *A. villadai* (Aves: Trochilidae) along the Mexican Transition Zone. Mitochondrial and nuclear DNA of individuals distributed in Mexico and southern United States (Fig. 3.12a) was sequenced and analyzed using maximum likelihood and Bayesian methods, and the historical demography, paleodistribution modeling, and niche divergence were also analyzed. The genetic divergence between the species was shallow, but the phylogenic tree (Fig. 3.12b) showed three independent lineages: one including populations of *A. violiceps* located north of the Transmexican Volcanic Belt, another including a mixture of *A. violiceps* and *A. viridifrons* south of the Transmexican Volcanic Belt, and a third with the populations of *A. villadai* distributed east of the Isthmus of Tehuantepec. The results were interpreted by the author as suggesting diversification by isolation with dispersal and habitat shifting in response to Pleistocene climatic fluctuations.

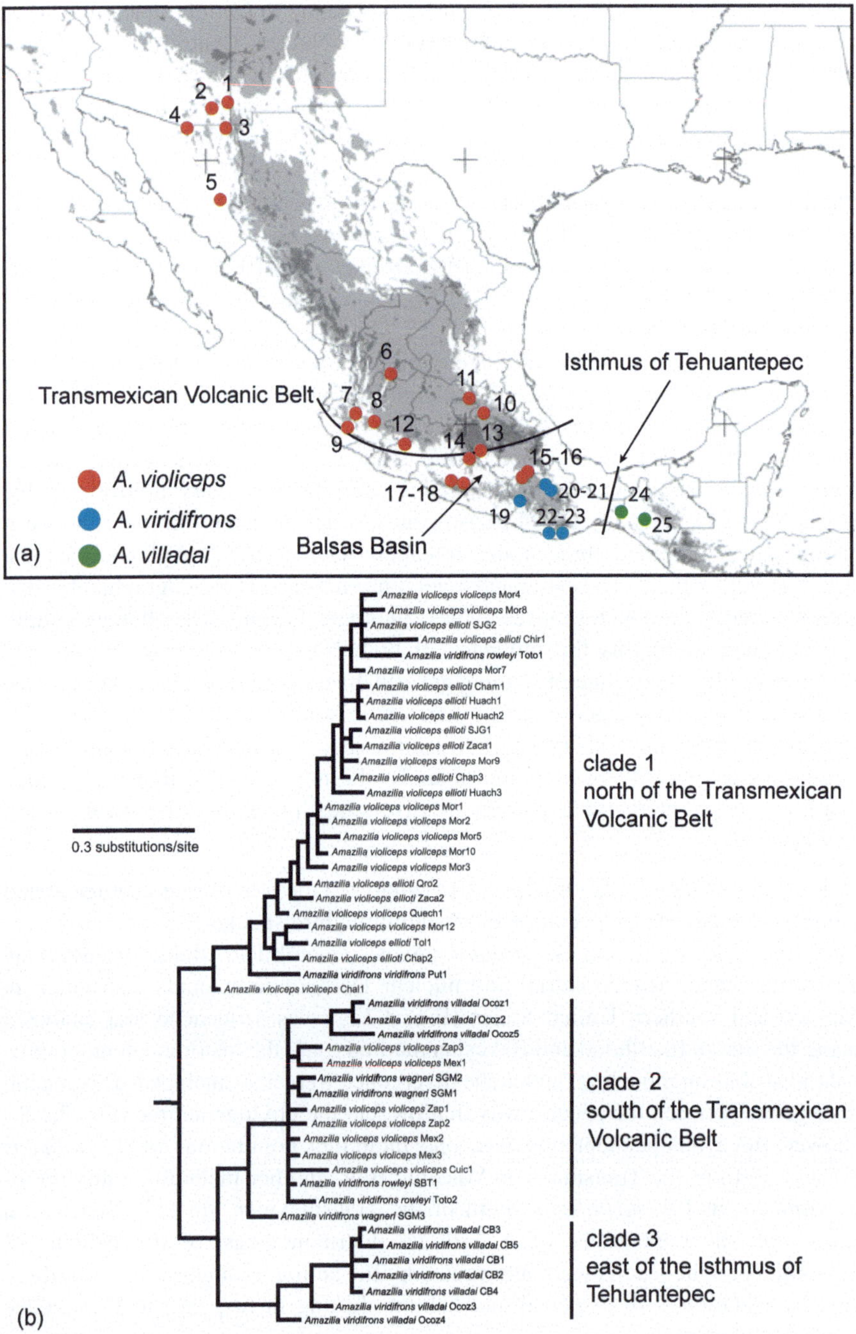

Fig. 3.12 Phylogeographic analysis of three species of *Amazilia* (Aves: Trochilidae) undertook by Rodríguez-Gómez and Ornelas (2015). (**a**) Sampling localities; (**b**) phylogenetic tree based on DNA sequence data

References

Abrahamovich A, Díaz NB, Morrone JJ (2004) Distributional patterns of the Neotropical and Andean species of the genus *Bombus* (Hymenoptera: Apidae). Acta Zool Mex 20:99–117

Agorreta A, Domínguez-Domínguez O, Reina RG, Miranda R, Bermingham E, Doadrio I (2013) Phylogenetic relationships and biogeography of *Pseudoxiphophorus* (Teleostei: Poeciliidae) based on mitochondrial and nuclear genes. Mol Phylogenet Evol 66:80–90

Almendra AL, Rogers DS, González-Cozatl FX (2014) Molecular phylogenetics of the *Handleyomys chapmani* complex in Mesoamerica. J Mammal 95:26–40

Andrés Hernández AR, Morrone JJ, Terrazas T, López Mata L (2006) Análisis de trazos de las especies mexicanas de *Rhus* subgénero *Lobadium* (Angiospermae: Anacardiaceae). Interciencia 31:900–904

Axelrod DI (1979) Age and origin of Sonoran desert vegetation. Occas Pap California Acad Sci 132:1–72

Ball GE (1968) Barriers and southward dispersal of the Holarctic boreo-montane elements of the family Carabidae in the mountains of Mexico. An Esc Nac Cienc Biol México 17:91–112

Brooks DR (2004) Reticulations in historical biogeography: the triumph of time over space in evolution. In: Lomolino MV, Heaney LR (eds) Frontiers of biogeography: new directions in the geography of nature. Sinauer Associates, Sunderland, MA, pp 111–124

Bryson RW Jr, García-Vázquez UO, Riddle BR (2011) Phylogeography of middle American gopher snakes: mixed responses to biogeographical barriers across the Mexican Transition Zone. J Biogeogr 38:1570–1584

Cabrera AL, Willink A (1973) Biogeografía de América Latina.. Monografía 13, Serie de Biología. OEA, Washington, DC

Castillo C, Reyes-Castillo P (1984) Biosistemática del género *Petrejoides* Kuwert (Coleoptera, Lamellicornia, Passalidae). Acta Zool Mex 4:1–84

Contreras-Medina R, Luna Vega I, Morrone JJ (2007) Gymnosperms and cladistic biogeography of the Mexican Transition Zone. Taxon 56:905–915

Corona AM, Toledo VH, Morrone JJ (2009) Track analysis of the Mexican species of Buprestidae (Coleoptera): testing the complex nature of the Mexican Transition Zone. J Biogeogr 36:1730–1738

Corral-Rosas V, Morrone JJ (2017) Analyzing the assembly of cenocrons in the Mexican Transition Zone through a time-sliced cladistic biogeographic analysis. Austr Syst Bot 29:489–501

Cox CL, Streicher JW, Sheehy CM, Campbell JA, Chippindale PT (2012) Patterns of genetic differentiation among populations of *Smilisca fodiens*. Herpetologica 68:226–235

Crisci JV, Katinas L (2009) Darwin, historical biogeography, and the importance of overcoming binary oppositions. J Biogeogr 36:1027–1032

Croizat L (1958) Panbiogeography, vol 1 and 2. Published by the Author, Caracas

Croizat L (1964) Space, time, form: the biological synthesis. Published by the Author, Caracas

Darlington JP (1957) Zoogeography: the geographical distribution of animals. Wiley, New York

Daza JM, Smith EN, Páez VP, Parkinson CL (2009) Complex evolution in the Neotropics: the origin and diversification of the widespread genus *Leptodeira* (Serpentes: Colubridae). Mol Phylogenet Evol 53:653–667

Daza JM, Castoe TA, Parkinson CL (2010) Using regional comparative phylogeographic data from snake lineages to infer historical processes in Middle America. Ecography 33:343–354

Dugès E (1902) Algo sobre la distribución geográfica de algunas aves. Mem Rev Soc Cient Antonio Alzate 18:44–46

Dunn ER (1931) The herpetological fauna of the Americas. Copeia 3:106–119

Escalante T, Rodríguez G, Morrone JJ (2004) The diversification of Nearctic mammals in the Mexican Transition Zone. Biol J Linn Soc 83:327–339

Escalante T, Rodríguez G, Morrone JJ (2005) Las provincias biogeográficas del componente Mexicano de Montaña desde la perspectiva de los mamíferos continentales. Rev Mex Biodivers 76:199–205

Escalante T, Rodríguez G, Cao N, Ebach MC, Morrone JJ (2007) Cladistic biogeographic analysis suggests an early Caribbean diversification in Mexico. Naturwissenschaften 94:561–565
Escalante T, Noguera-Urbano EA, Corona W (2018) Track analysis of the Nearctic region: identifying complex areas with mammals. J Zool Syst Evol Res 56:466–477
Escobar F, Halffter G, Arellano L (2007) From forest to pasture: an evaluation of the influence of environment and biogeography on the structure of dung beetle (Scarabaeinae) assemblages along three altitudinal gradients in the Neotropical region. Ecography 30:193–208
Espinosa Organista D, Ocegueda Cruz S, Aguilar Zúñiga C, Flores Villela O, Llorente-Bousquets J (2008) El conocimiento biogeográfico de las especies y su regionalización natural. In: Sarukhán J (ed) Capital natural de México, Conocimiento actual de la biodiversidad, vol I. Conabio, Mexico City, pp 33–65
Espinosa D, Llorente J, Morrone JJ (2006) Historical biogeographic patterns of the species of *Bursera* (Burseraceae) and their taxonomical implications. J Biogeogr 33:1945–1958
Estrada Sánchez I, García-Cruz J, Espejo-Serna A, López-Ortega G (2019) Identification of areas of endemism in the Mexican cloud forests based on the distribution of endemic epiphytic bromeliads and orchids. Phytotaxa 397:129–145
Evans HE (1966) A revision of the Mexican Central American spider wasps of the subfamily Pompilinae (Hymenoptera: Pompilidae). Mem Amer Entomol Soc 20:1–442
Flores-Villela O, Goyenechea I (2001) A comparison of hypotheses of historical biogeography for Mexico and Central America, or in search for the lost pattern. In: Johnson JD, Webb RG, Flores Villela O (eds) Mesoamerican herpetology: systematics, zoogeography, and conservation. The University of Texas at El Paso, El Paso, pp 171–181
Flores-Villela O, Martínez-Salazar EA (2009) Historical explanation of the origin of the herpetofauna of Mexico. Rev Mex Biodivers 80:817–833
García Díaz R, Cuevas Sánchez JA, Segura Ledesma S, Basurto Peña F (2015) Análisis panbiogeográfico de *Diospyros* spp. (Ebenaceae) en México. Rev Mex Cienc Agríc 6:187–200
García-Marmolejo G, Escalante T, Morrone JJ (2008) Establecimiento de prioridades para la conservación de mamíferos terrestres neotropicales de México. Mastozool Neotr 15:41–65
García-Navarrete PG, Morrone JJ (2018) Testing the biogeographical regionalization of the Mexican Transition Zone based on the distribution of Curculionidae (Coleoptera). Zootaxa 4530:1–99
Goldman EA, Moore RT (1945) The biotic provinces of Mexico. J Mammal 26:347–360
González-Ávila A, Contreras-Medina R, Espinosa D, Luna-Vega I (2017) Track analysis of the order Gomphales (Fungi: Basidiomycota) in Mexico. Phytotaxa 316:22–38
Gutiérrez-Ortega JS, Salinas-Rodríguez MM, Martínez JF, Molina-Freaner F, Pérez-Farrera MA, Vovides AP, Matsuki Y, Suyama Y, Ohsawa TA, Watano Y, Kajita T (2018) The phylogeography of the cycad genus *Dioon* (Zamiaceae) clarifies its Cenozoic expansion and diversification in the Mexican Transition Zone. Ann Bot 121:535–548
Gutiérrez-Velázquez A, Rojas-Soto O, Reyes-Castillo P, Halffter G (2013) The classic theory of Mexican Transition Zone revisited: the distributional congruence patterns of Passalidae (Coleoptera). Invertebr Syst 27:282–293
Halffter G (1962) Explicación preliminar de la distribución geográfica de los Scarabaeidae mexicanos. Acta Zool Mex 5:1–17
Halffter G (1964a) La entomofauna americana, ideas acerca de su origen y distribución. Folia Entomol Mex 6:1–108
Halffter G (1964b) Las regiones Neártica y Neotropical, desde el punto de vista de su entomofauna. Anais do II Congreso Latino-Americano de Zoología, São Paulo 1(1962):51–61
Halffter G (1965) Algunas ideas acerca de la zoogeografía de América. Rev Soc Mex Hist Nat 26:1–16
Halffter G (1968) La distribución de los insectos en la Zona de Transición Mexicana. Folia Entomol Mex 18–19:107–110
Halffter G (1972) Éléments anciens de l'entomofaune neotropicale: Ses implications biogéographiques. In: Biogeographie et Liasons Intercontinentales au cours du Mésozoique. 17me Congrés Internationale de Zoologie, Monte Carlo 1, pp 1–40

Halffter G (1974) Eléments anciens de l'entomofaune neotropicale: Ses implications biogéographiques. Quaest Entomol 10:223–262

Halffter G (1975) Éléments anciens de l'entomofaune néotropicale: Ses implications biogéographiques. Mem Mus Nat Hist Nat Nouv Sér Sér A Zool 88:114–145

Halffter G (1976) Distribución de los insectos en la Zona de Transición Mexicana: Relaciones con la entomofauna de Norteamérica. Folia Entomol Mex 35:1–64

Halffter G (1978) Un nuevo patrón de dispersión en la Zona de Transición Mexicana: El mesoamericano de montaña. Folia Entomol Mex 39–40:219–222

Halffter G (1987) Biogeography of the montane entomofauna of Mexico and Central America. Annu Rev Entomol 32:95–114

Halffter G (2003) Biogeografía de la entomofauna de montaña de México y América Central. In: Morrone JJ, Llorente Bousquets J (eds) Una perspectiva latinoamericana de la biogeografía. Las Prensas de Ciencias, UNAM, Mexico City, pp 87–97

Halffter G (2006) Biogeografía de la entomofauna de montaña de México y América Central. In: Morrone JJ, Llorente Bousquets J (eds) Componentes bióticos principales de la entomofauna mexicana. Las Prensas de Ciencias, UNAM, Mexico City, pp 1–21

Halffter G (2017) La Zona de Transición Mexicana y la megadiversidad de México: Del marco histórico a la riqueza actual. Dugesiana 24:78–89

Halffter G, Morrone JJ (2017) An analytical review of Halffter's Mexican Transition Zone, and its relevance for evolutionary biogeography, ecology and biogeographical regionalization. Zootaxa 4226:1–46

Halffter G, Favila ME, Arellano L (1995) Spatial distribution of three groups of Coleoptera along an altitudinal transect in the Mexican Transition Zone and its biogeographical implications. Elytron 9:151–185

Halffter G, Llorente-Bousquets J, Morrone JJ (2008a) La perspectiva biogeográfica histórica. In: Sarukhán J (ed) Capital natural de México, Conocimiento actual de la biodiversidad, vol I. Conabio, Mexico City, pp 67–86

Halffter G, Verdú JR, Márquez J, Moreno CE (2008b) Biogeographical analysis of Scarabaeinae and Geotrupinae along a transect in Central Mexico (Coleoptera, Scarabaeoidea). Fragm Entomol 40:273–322

Halffter G, Zunino M, Moctezuma V, Sánchez-Huerta JL (2019) The integration of the distributional patterns in the Mexican Transition Zone: phyletic, paleogeographic and ecological factors of a case study. Zootaxa 4586:1–34

Heilprin A (1887) The geographical and geological distribution of animals. International Scientific Series, New York

Hoffmann CC (1936) Relaciones zoogeográficas de los lepidópteros mexicanos. An Inst Biol UNAM 7:47–58

Huidobro L, Morrone JJ, Villalobos JL, Álvarez F (2006) Distributional patterns of freshwater taxa (fishes, crustaceans and plants) from the Mexican Transition Zone. J Biogeogr 33:731–741

Jeannel R (1942) La genese des faunes terrestres: Élements de biogéographie. Presses Universitaires de France, Paris

Kobelkowsky-Vidrio T, Ríos-Muñoz CA, Navarro-Sigüenza AG (2014) Biodiversity and biogeography of the avifauna of the Sierra Madre occidental, Mexico. Biodivers Conserv 23:2087–2105

Kohlmann B, Halffter G (1988) Cladistic and biogeographical analysis of *Ateuchus* (Coleoptera: Scarabaeidae) of Mexico and the United States. Folia Entomol Mex 74:109–130

Kohlmann B, Halffter G (1990) Reconstruction of a specific example of insect invasion waves: the cladistic analysis of *Canthon* (Coleoptera: Scarabaeidae) and related genera in North America. Quaest Entomol 26:1–28

Lanteri AA (1990) Systematic revision and cladistic analysis of *Phacepholis* horn (Coleoptera: Curculionidae). Southwest Entomol 15:179–204

Lieberman BS (2004) Range expansion, extinction, and biogeographic congruence: a deep time perspective. In: Lomolino MV, Heaney LR (eds) Frontiers of biogeography: new directions in the geography of nature. Sinauer Associates, Sunderland, MA, pp 111–124

Liebherr JK (1991) A general area cladogram for montane Mexico based on distributions in the Platynine genera *Elliptoleus* and *Calathus* (Coleoptera: Carabidae). Proc Entomol Soc Wash 93:390–406

Liebherr JK (1994a) Biogeographic patterns of montane Mexican and central American Carabidae (Coleoptera). Can Entomol 126:841–860

Liebherr JK (1994b) Identification of New World *Agonum*, review of the Mexican fauna, and description of *Incagonum*, new genus, from South America (Coleoptera: Carabidae: Platynini). J New York Entomol Soc 102:1–55

Llorente Bousquets J (1996) Biogeografía de artrópodos de México: ¿Hacia un nuevo enfoque? In: Llorente Bousquets J, García Aldrete AN, Gónzalez Soriano E (eds) Biodiversidad, taxonomía y biogeografía de artrópodos de México: Hacia una síntesis de su conocimiento. Universidad Nacional Autónoma de México, Mexico City, pp 41–56

Lobo JM (2007) Los "patrones de dispersión" de la fauna ibérica de Scarabaeinae (Coleoptera). In: Zunino M, Melic A (eds) Escarabajos, diversidad y conservación biológica: Ensayos en homenaje a Gonzalo Halffter,. Monografías 3er. Milenio M3M. Sociedad Entomológica Aragonesa, Zaragoza, pp 159–177

Lobo JM, Halffter G (2000) Biogeographical and ecological factors affecting the altitudinal variation of mountainous communities of coprophagous beetles (Coleoptera: Scarabaeoidea): a comparative study. Ann Entomol Soc Am 93:115–126

Luna Vega I, Alcántara Ayala O, Espinosa Organista D, Morrone JJ (1999) Historical relationships of the Mexican cloud forests: a preliminary vicariance model applying parsimony analysis of endemicity to vascular plant taxa. J Biogeogr 26:1299–1305

MacVean C, Schuster JC (1981) Altitudinal distribution of Passalid beetles (Coleoptera, Passalidae) and Pleistocene dispersal on the volcanic chain of northern Central America. Biotropica 13:29–38

Marshall CJ, Liebherr JK (2000) Cladistic biogeography of the Mexican Transition Zone. J Biogeogr 27:203–216

Martins UR (1971) Monografia do tribo Ibidionini (Coleop. Cerambycinae). Arq Zool São Paulo 16:1343–1508

Mateu J (1974) Sobre algunos linajes de carábidos boreo-montanos de México y sus relaciones con el poblamiento del Sistema Volcánico Transversal. Rev Soc Mex Hist Nat 35:181–224

Matthews EG (1966) A taxonomic and zoogeographic survey of the Scarabaeinae of the Antilles. Mem Amer Entomol Soc 21:1–134

Miguez-Gutiérrez A, Castillo J, Márquez J, Goyenechea I (2013) Biogeografía de la Zona de Transición Mexicana con base en un análisis de árboles reconciliados. Rev Mex Biodivers 84:215–224

Moctezuma V, Halffter G, Escobar F (2016) Response of copronecrophagous beetle communities to habitat disturbance in two mountains of the Mexican Transition Zone: influence of historical and ecological factors. J Inst Conserv 20:945–956

Morrone JJ (2003) Las ideas biogeográficas de Osvaldo Reig y el desarrollo del "dispersalismo" en América Latina. In: Morrone JJ, Llorente J (eds) Una perspectiva latinoamericana de la biogeografía. Las Prensas de Ciencias, UNAM, Mexico City, pp 69–74

Morrone JJ (2005) Hacia una síntesis biogeográfica de México. Rev Mex Biodivers 76:207–252

Morrone JJ (2006) Biogeographic areas and transition zones of Latin America and the Caribbean Islands based on panbiogeographic and cladistic analyses of the entomofauna. Annu Rev Entomol 51:467–494

Morrone JJ (2009) Evolutionary biogeography: an integrative approach with case studies. Columbia University Press, New York

Morrone JJ (2010) Fundamental biogeographic patterns across the Mexican Transition Zone: an evolutionary approach. Ecography 33:355–361

Morrone JJ (2011) La teoría biogeográfica de Florentino Ameghino y el carácter episódico de la evolución geobiótica de los mamíferos terrestres de América del Sur. Asoc Paleontol Argent Publ Espec 12:81–89

Morrone JJ (2014a) Cladistic biogeography of the Neotropical region: identifying the main events in the diversification of the terrestrial biota. Cladistics 30:202–214
Morrone JJ (2014b) Biogeographical regionalisation of the Neotropical region. Zootaxa 3782:1–110
Morrone JJ (2014c) On biotas and their names. Syst Biodivers 12:386–392
Morrone JJ (2015a) Halffter's Mexican Transition Zone (1962–2014), cenocrons and evolutionary biogeography. J Zool Syst Evol Res 53:249–257
Morrone JJ (2015b) Track analysis beyond panbiogeography. J Biogeogr 42:413–425
Morrone JJ (2015c) Biogeographical regionalisation of the world: a reappraisal. Austr Syst Bot 28:81–90
Morrone JJ (2017) Neotropical biogeography: regionalization and evolution. CRC Press, Taylor and Francis, Boca Raton, FL
Morrone JJ, Escalante T (2016) Introducción a la biogeografía. Las Prensas de Ciencias, UNAM, Mexico City
Morrone JJ, Gutiérrez A (2005) Do fleas (Insecta: Siphonaptera) parallel their mammal host diversification in the Mexican Transition Zone? J Biogeogr 32:1315–1325
Morrone JJ, Márquez J (2001) Halffter's Mexican Transition Zone, beetle generalised tracks, and geographical homology. J Biogeogr 28:635–650
Munguía-Lino G, Escalante T, Morrone JJ, Rodríguez A (2016) Areas of endemism of the north American species of Tigridieae (Iridaceae). Austr Syst Bot 29:142–156
Navarro-Sigüenza A, Peterson AT, Nyari A, García-Derás G, García Moreno J (2008) Phylogeography of the *Buarremon* brush-finch complex (Aves, Emberizidae) in Mesoamerica. Mol Phylogenet Evol 47:21–35
Nolasco-Soto J, González-Astorga J, Espinosa de los Monteros A, Galente-Patiño E, Favila ME (2017) Phylogeographic structure of *Canthon cyanellus* (Coleoptera: Scarabaeidae), a Neotropical dung beetle in the Mexican Transition Zone: insights on its origin and the impacts of Pleistocene fluctuations on population dynamics. Mol Phylogenet Evol 109:180–190
Ortega J, Arita HT (1998) Neotropical-Nearctic limits in middle America as determined by distributions of bats. J Mammal 79:772–781
Pinilla-Buitrago GE, Escalante T, Gutiérrez-Velázquez A, Reyes-Castillo P, Rojas-Soto OR (2018) Areas of endemism persist through time: a paleoclimatic analysis in the Mexican Transition Zone. J Biogeogr 45:952–961
Rapoport EH (1968) Algunos problemas biogeográficos del Nuevo Mundo con especial referencia a la región Neotropical. In: Delamare Debouteville C, Rapoport EH (eds) Biologie de l'Amerique Australe, vol 4. Editions du Centre National de la Recherche Scientifique, Paris, pp 55–110
Rapoport EH (1971) The nearctic-neotropical frontiers. Proceedings of the XIII international congress of entomology, Nauka, Leningrad, pp 190–191
Reichardt H (1977) A sinopsis of the genera of Neotropical Carabidae (Insecta: Coleoptera). Quaest Entomol 13:346–493
Reig OA (1962) Las integraciones cenogenéticas en el desarrollo de la fauna de vertebrados tetrápodos de América del Sur. Ameghiniana 2:131–140
Reig OA (1968) Peuplement de vertébrés tétrapodes de l'Amerique de Sud. In: Delamare Debouteville C, Rapoport EH (eds) Biologie de l'Amerique Australe, vol 4. Editions du Centre National de la Recherche Scientifique, Paris, pp 215–260
Reig OA (1981) Teoría del origen y desarrollo de la fauna de mamíferos de América del Sur. Museo Municipal de Ciencias Naturales Lorenzo Scaglia, Mar del Plata
Reyes-Castillo P (1970) Coleoptera, Passalidae: Morfología y división en grandes grupos: Géneros americanos. Folia Entomol Mex 20–22:1–240
Reyes-Castillo P (2003) Las ideas biogeográficas de Gonzalo Halffter: Importancia e impacto. In: Morrone JJ, Llorente Bousquets J (eds) Una perspectiva latinoamericana de la biogeografía. Las Prensas de Ciencias, UNAM, Mexico City, pp 99–108

Reyes-Castillo P, Halffter G (1978) Análisis de la distribución de la tribu Proculini (Coleoptera, Passalidae). Folia Entomol Mex 39–40:222–226
Riddle BR, Dawson MN, Hadly EA, Hafner DJ, Hickerson MJ, Mantooth SJ, Yoder AD (2008) The role of molecular genetics in sculpting the future of integrative biogeography. Prog Phys Geogr 32:173–202
Rodríguez A, Castro-Castro A, Vargas-Amado G, Vargas-Ponce O, Zamora-Tavares P, González-Gallegos J, Carrillo-Reyes P, Anguiano-Constante M, Carrasco-Ortiz M, García-Martínez M, Gutiérrez-Rodríguez B, Aragón-Parada J, Valdés-Ibarra C, Munguía-Lino G (2018) Richness, geographic distribution patterns, and areas of endemism of selected angiosperm groups in Mexico. J Syst Evol 56:537–549
Rodríguez-Gómez F, Ornelas JF (2015) At the passing gate: past introgression in the process of species formation between *Amazilia violiceps* and *A. viridifrons* hummingbirds along the Mexican Transition Zone. J Biogeogr 42:1305–1318
Roig-Juñent SA (2005) Las ideas biogeográficas de René Jeannel y su impacto en el conocimiento de la biogeografía de América del Sur. In: Llorente Bousquets J, Morrone JJ (eds) Regionalización biogeográfica en Iberoamérica y tópicos afines: Primeras Jornadas Biogeográficas de la Red Iberoamericana de Biogeografía y Entomología Sistemática (RIBES XII.I-CYTED). Las Prensas de Ciencias, UNAM, Mexico City, pp 55–66
Roig-Juñent SA, Tognelli MF, Morrone JJ (2008) Aspectos biogeográficos de los insectos de la Argentina. In: Claps LE, Debandi G, Roig-Juñent S (eds) Biodiversidad de artrópodos argentinos, vol 2. Sociedad Entomológica Argentina, San Miguel de Tucumán, pp 11–29
Roig-Juñent SA, Griotti M, Domínguez MC, Agrain FA, Campos-Soldini P, Carrara R, Cheli G, Fernández-Campón F, Flores GE, Katinas L, Muzón JR, Neita-Moreno JC, Pessacq P, San Blas G, Scheibler EE, Crisci JV (2018) The Patagonian steppe biogeographic province: Andean region or south American transition zone? Zool Scr 47:623–629
Rosas MV, Morrone JJ, del Río MG, Lanteri AA (2011a) Phylogenetic analysis of the *Pantomorus-Naupactus* complex (Coleoptera: Curculionidae: Entiminae) from north and Central America. Zootaxa 2780:1–19
Rosas MV, del Río MG, Lanteri AA, Morrone JJ (2011b) Track analysis of the north and central American species of the *Pantomorus-Naupactus* complex (Coleoptera: Curculionidae). J Zool Syst Evol Res 49:309–314
Ryan RM (1963) The biotic provinces of Central America. Acta Zool Mex 6:1–55
Rzedowski J (1978) La vegetación de México. Editorial Limusa, Mexico City
Sánchez-González L, Navarro-Sigüenza AG (2009) History meets ecology: a geographical analysis of ecological restriction in the Neotropical humid montane forests avifaunas. Divers Distrib 15:1–11
Sanmartín I, Ronquist F (2004) Southern hemisphere biogeography inferred by event-based models: plant versus animal patterns. Syst Biol 53:216–243
Santiago-Alvarado M (2019) Evaluación de la validez de las provincias de la Zona de Transición Mexicana. Ms. Sci. dissertation, FES Zaragoza, UNAM, Mexico City
Savage JM (1966) The origins and history of the central American herpetofauna. Copeia 1966:719–766
Savage JM (1982) The enigma of the central American herpetofauna: dispersals or vicariance? Ann Missouri Bot Gard 69:464–547
Smith HM (1941) Las provincias bióticas de México, según la distribución geográfica de las lagartijas del género *Sceloporus*. An Esc Nac Cienc Biol 2:103–110
Smith HM (1949) Herpetogeny in Mexico and Guatemala. Ann Assoc Am Geogr 39:219–238
Toledo VH, Corona AM, Morrone JJ (2007) Track analysis of the Mexican species of Cerambycidae (Insecta, Coleoptera). Rev Bras Entomol 51:131–137
Udvardy MDF (1969) Dynamic zoogeography with special reference to land animals. Van Nostrand Reinhold, New York

Vázquez-Miranda H, Navarro-Sigüenza A, Omland K (2009) Phylogeography of the rufous-naped wren (*Campylorhynchus rufinucha*): speciation and hybridization in Mesoamerica. Auk 126:765–768

Vivó JA (1943) Los límites biogeográficos de América y la zona cultural mesoamericana. Rev Geogr 3:109–131

Wallace AR (1876) The geographical distribution of animals. Macmillan, London

Zamora-Tavares MP, Martínez M, Magallón S, Guzmán-Dávalos L, Vargas-Ponce O (2016) *Physalis* and physaloids: a recent and complex evolutionary history. Mol Phylogenet Evol 100:41–50

Zunino M (1985) Gli Scarabaeoidea coprofagi dell'area Euromediterranea: Relazione filetiche e biogeografiche. Atti XIV Congr. Naz. Ital. Ent., Palermo, pp 321–325

Zunino M (2007) Latinoamérica ante las ciencias de la naturaleza y del medio ambiente: Materiales para una reflexión histórica. Acta Zool Mex 23:181–190

Zunino M, Halffter G (1988) Análisis taxonómico, ecológico y biogeográfico de un grupo americano de *Onthophagus* (Coleoptera: Scarabaeidae). Mus Reg Sci Nat Mon 9:1–211

Zunino M, Zullini A (2003) Biogeografía: La dimensión espacial de la evolución. Fondo de Cultura Económica, Mexico City

Chapter 4
Biogeographic Regionalization of the Mexican Transition Zone

Order and simplification are the first steps toward the mastery of a subject; the actual enemy is the unknown.

Thomas Mann (1924), The Magic Mountain

Abstract The Mexican Transition Zone is the area where the Neotropical and Nearctic regions overlap. In its strict sense followed in this book, it corresponds to the moderate- to high-elevation highlands of Mexico, Guatemala, Honduras, El Salvador, and Nicaragua. This area is considered a transition zone between the Nearctic and Neotropical regions, so from the perspective of biogeographic regionalization, it should be assigned simultaneously to both regions. Within the Mexican Transition Zone, I recognize 5 biogeographic provinces and 22 districts. The Sierra Madre Occidental province is situated in western Mexico at elevations between 200 and 3000 m, with most of the area above 2000 m; it includes the Apachian and Durangoan districts. The Sierra Madre Oriental province is situated in eastern Mexico at elevations above 1500 m; it includes two subprovinces: Austral-Oriental (with the Saltillo-Parras and Potosí districts) and Hidalgoan (with the Sierra Gorda and Zacualtipán districts). The Transmexican Volcanic Belt province is situated in central Mexico, at elevations above 1800 m; it includes two subprovinces: West (with the Otomí and Tarascan districts) and East (with the Aztec and Orizaba-Zempoaltepec districts). The Sierra Madre del Sur province comprises south central Mexico at elevations above 1000 m; it includes three subprovinces: Western Sierra Madre del Sur (with the Jaliscian and Jaliscian-Manantlán districts), Central Sierra Madre del Sur subprovince (Michoacán district), and Eastern Sierra Madre del Sur (with the Guerreran and Oaxacan Highlands districts). The Chiapas Highlands province comprises southern Mexico, Guatemala, Honduras, El Salvador, and Nicaragua, basically corresponding to the Sierra Madre de Chiapas, from 500 to 2000 m altitude; it includes the Sierra Madrean, Comitanian, Lacandonian, Soconusco, Guatemalan Highland, and Nicaraguan Montane districts.

J. J. Morrone, *The Mexican Transition Zone*, https://doi.org/10.1007/978-3-030-47917-6_4

4.1 Introduction

The Mexican Transition Zone in its strict sense corresponds to the moderate- to high-elevation highlands of Mexico, Guatemala, Honduras, El Salvador, and Nicaragua (Morrone 2010, 2014a, 2017a, 2019). These highlands correspond to five physiographic units (Fig. 4.1) (Challenger 1988; Ferrusquía-Villafranca 1998):

Sierra Madre Occidental It represents the largest mountainous system in Mexico. It is parallel to the Pacific Ocean coast from the Mexican-US border to the states of Nayarit and Jalisco, where it connects with the Transmexican Volcanic Belt. It represents the southern extension of the Rocky Mountains of the United States and separates the Northwest Coastal Plain from the Mexican Highlands. It is 1500 km long and its highest altitude is over 3000 m.

Sierra Madre Oriental It begins in the state of Nuevo León and runs south to southeast, connecting with the Transmexican Volcanic Belt in Puebla and Veracruz. It separates the Mexican Plateau from the Northeast Coastal Plain. It is about 1350 km long and has some important elevations, such as Cerro Potosí (3650 m) in

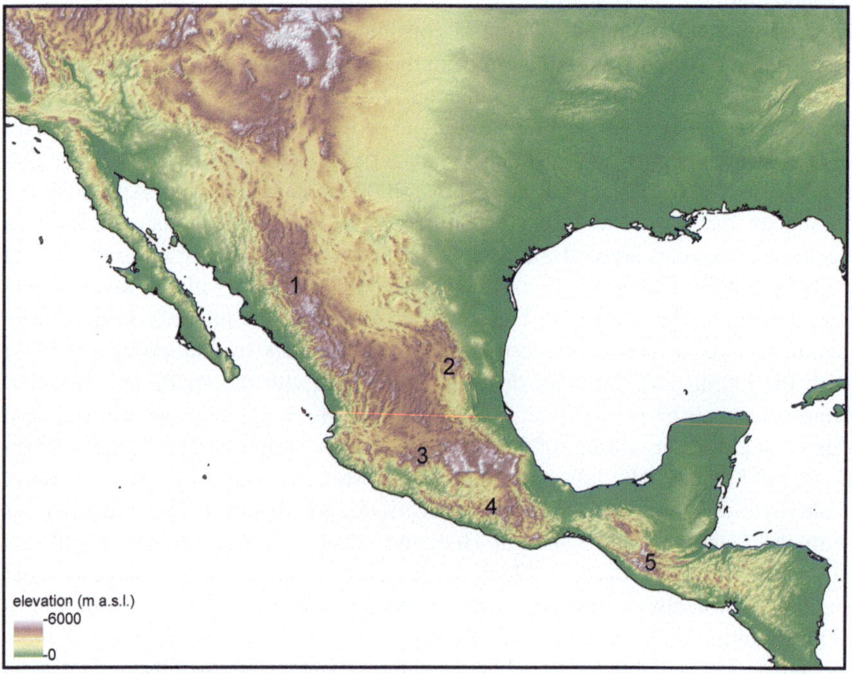

Fig. 4.1 Main physiographic units of the Mexican Transition Zone. (1) Sierra Madre Occidental, (2) Sierra Madre Oriental, (3) Transmexican Volcanic Belt, (4) Sierra Madre del Sur (5) Central American Nucleus

Nuevo León and Cerro de San Antonio Peña Nevada (3450 m) between Nuevo León and Tamaulipas.

Transmexican Volcanic Belt This system crosses Mexico in an east-west direction, approximately along the parallels of 19 and 20°N. It is about 1000 km long and 50–150 km wide and separates the Mexican Plateau from the Balsas Basin. It includes the major elevations in the country, namely, the volcanoes Pico de Orizaba (5650 m), Popocatépetl (5450 m), Ixtaccíhuatl (5280 m), Nevado de Toluca (4560 m), La Malinche (4460 m), Nevado de Colima (4340 m), Tancítaro (4160 m), Tláloc (4150 m), and Cofre de Perote (4090 m), although its average height is 2000 m.

Sierra Madre del Sur It runs northwest to southeast from the state of Jalisco to the Isthmus of Tehuantepec, very close to the coast of the Pacific Ocean. It has about 1100 km in length and its average width is 120 km. Its altitudes are greater than 1000 m and are interrupted by the valleys of the Armería, Balsas, Papagayo, and Verde rivers. The lowest elevations are about 300–500 m high, and the maximum reaches 3700 m, at Cerro Nube, Oaxaca.

Central American Nucleus It consists of highlands of Central America in Guatemala, Honduras, El Salvador, and Nicaragua and two northern extensions in southern Mexico. The Sierra Madre de Chiapas runs along the Pacific coast of Chiapas and penetrates the Isthmus of Tehuantepec, reaching its highest altitude in Tacaná (4026 m), in the Mexican-Guatemalan border. The Central Massif of Chiapas also runs parallel to the Pacific Ocean coast, but further away from it. It reaches its highest altitude in the San Cristóbal area (2860 m).

Wallace (1876) originally recognized the Mexican subregion, which he assigned to the Neotropical region (Fig. 4.2). Darlington (1957) described the Central American-Mexican Transition Zone for the area of overlap of the Nearctic and Neotropical regions. The current understanding of the Mexican Transition Zone began with the contributions of Halffter (1962, 1964, 1965, 1974, 1976, 1978, 1987). In his contributions, Halffter has been reluctant to provide a biogeographic regionalization, sharing the view of other evolutionary biogeographers, for example, Mayr (1946), who stated:

> Eventually it was realized that the whole method of approach—*Fragestellung*—of this essentially static zoogeography was wrong. Instead of thinking of fixed regions, it is necessary to think of fluid faunas. (Mayr 1946, p. 5)

This view means that the distributional patterns of plants, animals, and other organisms are so complex that it is impossible to achieve a regionalization. In contrast, I believe that whenever biogeographers, macroecologists, evolutionary biologists, systematists, and conservationists communicate their ideas they need to refer to a biogeographic regionalization (Morrone 2017a, 2018). For example, when we see a reference to the Neotropical region or to the Sierra Madre Oriental province, immediately we may get a graphical image that synthesizes the delimitation of these

Fig. 4.2 Biogeographic subregions of the Neotropical region according to Wallace (1876)

areas, the taxa that characterize them, their vegetation, and the set of physiographic and climatic conditions. Additionally, the recognition of natural biogeographic units in a regionalization may help document efficiently biodiversity patterns in conservation biogeography. For these reasons, I think that a biogeographic regionalization of the Mexican Transition Zone has great relevance.

4.2 Mexican Transition Zone

Mexican subregion Wallace 1876, 78 (zoogeographic regionalization); Heilprin 1887, 80 (zoogeographic regionalization); Bartholomew et al. 1911, 9 (zoogeographic regionalization); Mello-Leitão 1937, 222 (zoogeographic regionalization).
Aztec province Engler 1882, 345 (phytogeographic regionalization).
Guatemalan province Engler 1882, 345 (phytogeographic regionalization).
Mexican Highlands region Engler 1882, 345 (phytogeographic regionalization).
Central American subarea (in part) Clarke 1892, 381 (zoogeographic regionalization).
Sonoran region (in part) Matthew 1915, 186 (zoogeographic regionalization); Lydekker 1896, 135 (zoogeographic regionalization).
Central American subregion (in part) Sclater and Sclater 1899, 65 (zoogeographic regionalization).
Central American-Mexican Transition Zone Darlington 1957, 456 (biotic evolution).
Extratropical Highlands kingdom West 1964, 365 (zoogeographic regionalization).
Tropical Highlands kingdom West 1964, 365 (zoogeographic regionalization).
Mexican Transition Zone Halffter 1965, 4 (biotic evolution); Halffter 1974, 229 (biotic evolution); Halffter 1976, 13 (biotic evolution); Halffter 1978, 219 (biotic evolution); Castillo and Reyes-Castillo 1984, 72 (biotic evolution); Halffter 1987, 95 (biotic evolution); Kohlmann and Halffter 1988, 112 (biotic evolution); Zunino and Halffter 1988, 175 (biotic evolution); Kohlmann and Halffter 1990, 1 (biotic evolution); Llorente Bousquets 1996, 50 (biotic evolution); Morón 1996, 323 (biotic evolution); Ortega and Arita 1998, 777 (zoogeographic regionalization); Marshall and Liebherr 2000, 203 (cladistic biogeography); Morrone and Márquez 2001, 636 (track analysis); Márquez and Morrone 2003, 23 (track analysis); Escalante et al. 2004, 327 (track analysis); Morrone 2004, 155 (track analysis); Corona and Morrone 2005, 37 (track analysis); Morrone 2005, 234 (biogeographic regionalization); Morrone and Gutiérrez 2005, 1315 (track analysis); Morrone 2006, 475 (biogeographic regionalization); Morrone and Llorente Bousquets 2006, 1021 (biogeographic regionalization); Contreras-Medina et al. 2007a, 905 (cladistic biogeography); Escalante et al. 2007a, 562 (cladistic biogeography); Mariño-Pérez et al. 2007, 80 (track analysis); Halffter et al. 2008, 69 (biotic evolution); Corona et al. 2009, 1731 (track analysis); Escalante et al. 2009, 473 (endemicity analysis); Morrone 2010, 355 (biotic evolution); Coulleri

and Ferrucci 2012, 105 (track analysis); Mercado-Salas et al. 2012, 459 (track analysis); Gutiérrez-Velázquez et al. 2013, 282 (parsimony analysis of endemicity); Miguez-Gutiérrez et al. 2013, 216 (cladistic biogeography); Ruiz-Sánchez and Specht 2013, 1337 (biotic evolution); Lamas et al. 2014, 955 (cladistic biogeography); Morrone 2014a, 27 (biogeographic regionalization); 2014b, 203 (cladistic biogeography); García Díaz et al. 2015, 195 (track analysis); Klassa and Santos 2015, 520 (endemicity analysis); Rodríguez-Gómez and Ornelas 2015, 1305 (phylogeography); Cuellar-Martínez and Sosa 2016, 688 (species richness and endemicity analyses); Munguía-Lino et al. 2016, 150 (endemicity analysis); Zamora-Tavares et al. 2016, 44 (ancestral area analysis); Corral-Rosas and Morrone 2017, 489 (cladistic biogeography); Ferro et al. 2017, 2146 (biotic evolution); Gómez-Ortiz et al. 2017, 121 (species richness); González-Ávila et al. 2017, 27 (track analysis); Halffter and Morrone 2017, 78 (biotic evolution); Halffter and Morrone 2017, 2 (biotic evolution); Morrone 2017a, 49 (biogeographic regionalization and biotic evolution); Morrone et al. 2017, 277 (map); Nolasco-Soto et al. 2017, 180 (phylogeography); Arriaga-Jiménez et al. 2018, 2 (diversity and richness); Beron 2018 (zoogeographic regionalization); García-Navarrete and Morrone 2018, 3 (parsimony analysis of endemicity); Gutiérrez-Ortega et al. 2018, 536 (phylogeography); Pinilla-Buitrago et al. 2018, 952 (parsimony analysis of endemicity); Rodríguez et al. 2018, 537 (species richness and endemicity); Aragón-Parada et al. 2019, 2 (diversity and richness); Halffter et al. 2019, 2 (biotic evolution); Morrone 2019, 31 (biogeographic regionalization).

Mountain Mesoamerican province Cabrera and Willink 1973, 32 (biogeographic regionalization); Huber and Riina 2003, 262 (phytogeographic glossary).

Mountain Mesoamerican region Rzedowski 1978, 101 (phytogeographic regionalization); Luna Vega et al. 1999, 1301 (parsimony analysis of endemicity).

Central American transition zone Müller 1986, 21 zoogeographic regionalization).

Megamexico 3 Rzedowski 1991, 11 (biotic evolution).

Central America bioregion (in part) Dinerstein et al. 1995, map (ecological regionalization).

Mexican Mountains kingdom Huber and Riina 2003, 167 (phytogeographic glossary).

Mexican Mountain component Morrone and Márquez 2003, 219 (track analysis); Escalante et al. 2005, 199 (parsimony analysis of endemicity).

Madrean Highlands area Porzecanski and Cracraft 2005, 266 (parsimony analysis of endemicity).

Mexican Mountain Transition Zone Espinosa Organista et al. 2008, 54 (biogeographic regionalization).

Biotic Mesoamerica Ríos-Muñoz 2013, 1027 (biotic evolution).

Nearctic-Neotropical Transition Zone Ficetola et al. 2017, 1 (zoogeographic regionalization).

The Mexican Transition Zone was defined by Halffter (e.g., 1965, 1976, 1978, 1987), but it was not incorporated formally to a scheme of biogeographic regionalization. There are several authors who have recognized areas that are equivalent to the Mexican Transition Zone, at least in the strict sense (Morrone 2006, 2014a, 2017a, 2019). For example, Cabrera and Willink (1973) described the Mountain Mesoamerican province (Fig. 4.3) in their biogeographic regionalization of Latin America and the Caribbean. This province corresponds to the highlands of Central America and Mexico, generally above 1000 m of altitude, and is characterized by species of *Quercus* (Fagaceae), *Pinus* (Pinaceae), and some Poaceae. From the zoogeographic viewpoint, these authors followed Halffter (1964) in considering the fauna as transitional, with vertebrates being mostly Nearctic and insects and other invertebrates being of Neotropical origin.

In his phytogeographic regionalization of Mexico, Rzedowski (1978) characterized the Mountain Mesoamerican region (Fig. 4.4), which he considered to belong simultaneously to the Holarctic and Neotropical kingdoms. In contrast to the zoogeographic regionalization, Rzedowski (1978) postulated that a broad transitional zone encompassing the mountain areas was a better representation of the plants distributed in Mexico. The rest of the country was assigned to the Neotropical kingdom, with the exception of the northern part of the Baja California Peninsula, which Rzedowski (1978) assigned to the Nearctic kingdom. Megamexico 3 (Rzedowski 1991) and Biotic Mesoamerica (Ríos-Muñoz 2013) are also equivalent to the Mexican Transition Zone.

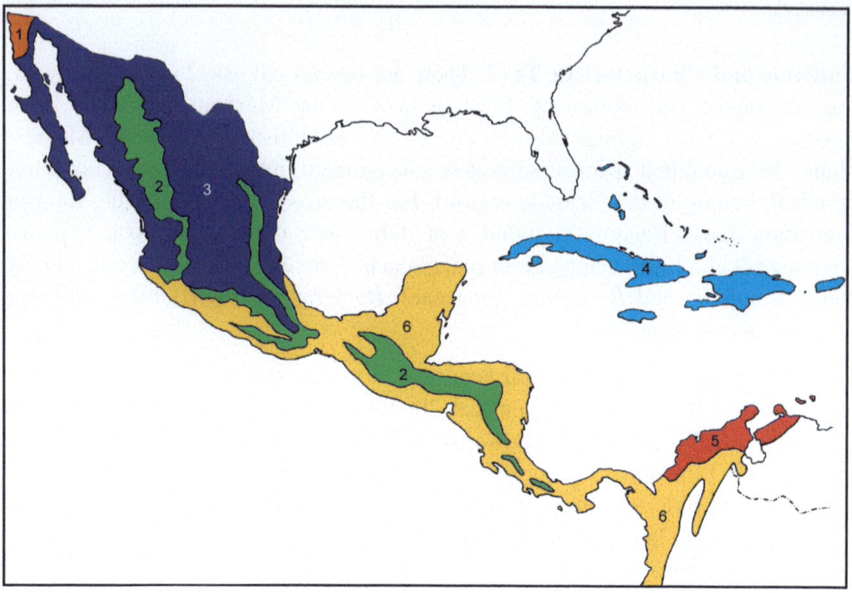

Fig. 4.3 Biogeographic provinces of Mexico and Central America (modified from Cabrera and Willink 1973). (1) Mountain Forest, (2) Mountain Mesoamerican, (3) Mexican Xerophyllous, (4) Caribbean, (5) Guajira, (6) Pacific

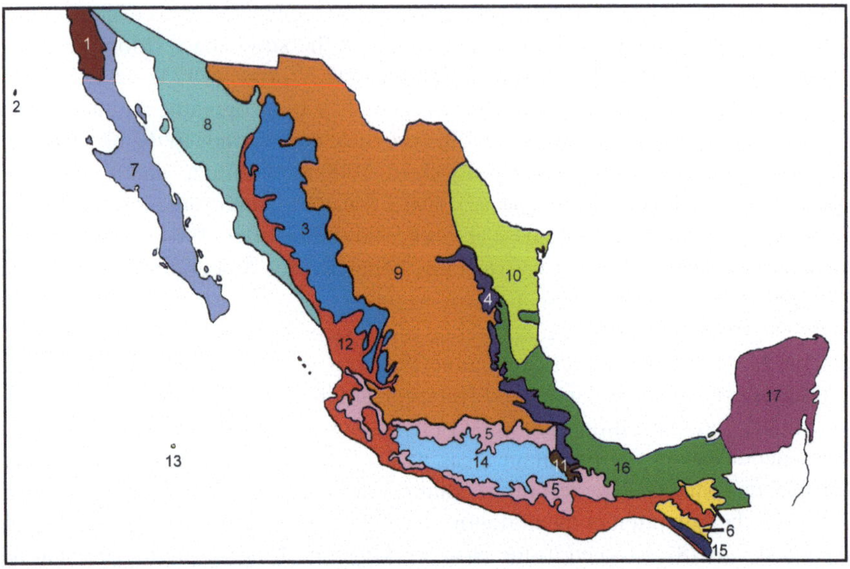

Fig. 4.4 Biogeographic provinces of Mexico (modified from Rzedowski 1978). (1) California, (2) Guadalupe Island, (3) Sierra Madre Occidental, (4) Sierra Madre Oriental, (5) Meridional Mountains, (6) Transisthmic Mountains, (7) Baja California, (8) Northwestern Coastal Plains, (9) Plateau, (10) Northeastern Coastal Plains, (11) Tehuacán-Cuicatlán Valley. (12) Pacific coast, (13) Revillagigedo Islands, (14) Balsas Basin, (15) Soconusco, (16) Mexican Gulf, (17) Yucatán Peninsula

Endemic and Characteristic Taxa There are several published studies analyzing the geographic distribution of different taxa in the Mexican Transition Zone. Rzedowski (1991) highlighted the richness and endemism of species of Mexican plants. He found that species richness is concentrated in the arid areas (considered herein to belong to the Nearctic region). For the areas belonging to the Mexican Transition Zone, Rzedowski found a predominance of Nearctic taxa. Espinosa Organista et al. (2008) mapped four plant taxa that are endemic to the area, namely, *Pinus leiophylla* and *P. teocote* (Pinaceae), *Bursera roseana* (Burseraceae), and *Quercus rugosa* (Fagaceae).

Vertebrate taxa include several genera of Rodentia (Mammalia), with Nearctic affinities, e.g., *Habromys* (Carleton et al. 2002; León-Paniagua et al. 2007), *Microtus* (Conroy et al. 2001), *Neotoma* (Edwards and Bradley 2002a, b), *Peromyscus* (Harris et al. 2000; Tiemann-Boege et al. 2000), and *Reithrodontomys* (Arellano et al. 2005). Among birds, the study of the *Chlorospingus ophthalmicus* species complex (Thraupidae) by García-Moreno et al. (2004) exemplifies the diversification of a taxon of Neotropical affinity.

Insects are among the best-studied taxa of the Mexican Transition Zone, including the classical contributions of Halffter (1964, 1965, 1974, 1976, 1978, 1987) on scarabaeid beetles. Castillo and Reyes-Castillo (1984) analyzed the species of *Petrejoides* (Coleoptera: Passalidae), finding different lineage characteristics of the Mexican Transition Zone. Kohlmann and Halffter (1988, 1990) analyzed the biotic evolution of the genera *Ateuchus* and *Canthon* (Coleoptera: Scarabaeidae), based on the phylogenetic relationships of their species. Liebherr (1991, 1994a, b) analyzed several taxa of Carabidae (Coleoptera), mapping distributional data of their species and classifying them according to Halffter's patterns. For example, species of the genus *Elliptoleus* illustrate the Nearctic pattern, whereas those of the *Platynus degallieri* species group illustrate the Mountain Mesoamerican pattern (Liebherr 1994a).

The arachnofauna of the Mexican Transition Zone is particularly rich in endemic species (Beron 2018), with all the arachnid orders represented. The most important are Scorpiones, Opiliones, Schizomida, and Araneae.

Vegetation Considering the main biomes that have been recognized for Mexico by Villaseñor and Ortiz (2013), the Mexican Transition Zone is characterized by both temperate and mountain humid forests. Temperate forests are found in the temperate and semihumid portions of the mountains from seal level to 3500 m, whereas the humid forests are concentrated in the mountains where the winds coming from the Pacific and Atlantic Oceans provide humidity. The latter are especially abundant in the Atlantic slope of the Sierra Madre Oriental, are more scattered in the Sierra Madre Occidental, and are scarce in the Transmexican Volcanic Belt. The distributional map of *Pinus* forests (Perry et al. 2000) matches closely the Mexican Transition Zone.

Biotic Relationships Based on its transitional nature, the relationships of the Mexican Transition Zone lie with the biotas of the Nearctic and Neotropical regions. Rzedowski (1963, 1978, 1991) suggested that the flora of the cloud forests of the Mexican Transition Zone comprises basically three elements: Nearctic or temperate taxa, particularly represented by canopy trees; Neotropical or tropical taxa, particularly represented by herbs, epiphytes, and shrubs; and endemic taxa, which are poorly represented at the generic level but very significant if species are considered. Halffter and Morrone (2017) characterized the cenocrons that contributed to the biotic assembly of the Mexican Transition Zone (see Chap. 5).

Regionalization The Mexican Transition Zone (Fig. 4.5) comprises the Sierra Madre Occidental, Sierra Madre Oriental, Transmexican Volcanic Belt, Sierra Madre del Sur, and Chiapas Highlands biogeographic provinces (Morrone 2014a, 2017a, 2019).

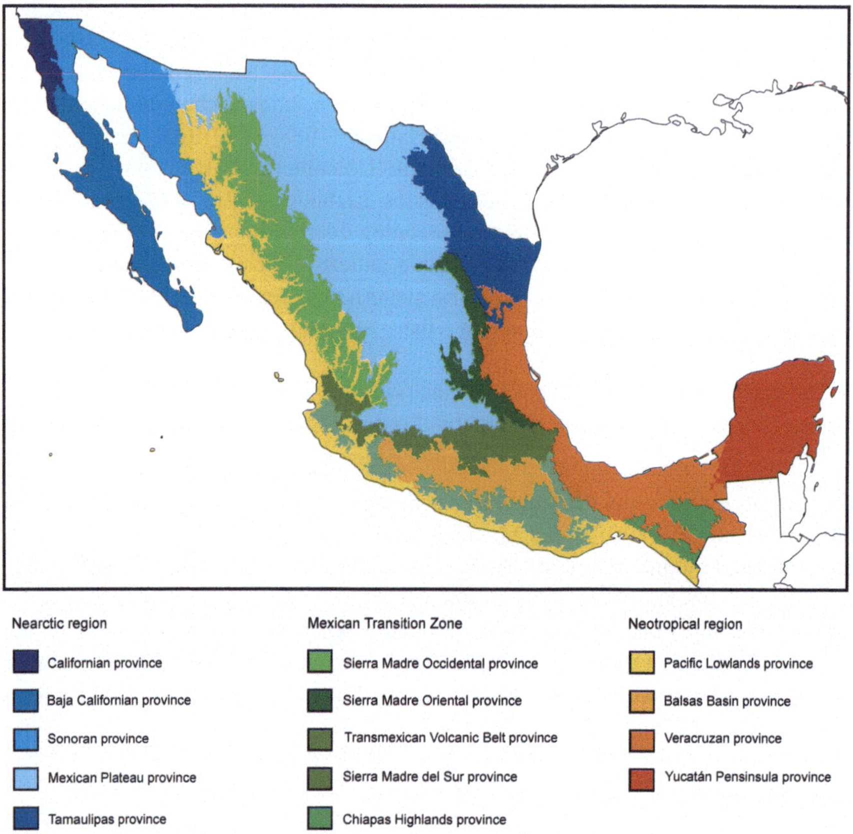

Fig. 4.5 Biogeographic provinces of Mexico (modified from Morrone et al. 2017)

4.3 Sierra Madre Occidental Province

Sierra Madre Occidental province Goldman and Moore 1945, 253 (zoogeographic regionalization); Moore 1945, 218 (zoogeographic regionalization); Stuart 1964, 350 (zoogeographic regionalization); Rzedowski 1978, 102 (phytogeographic regionalization); Casas-Andreu and Reyna-Trujillo 1990, map (zoogeographic regionalization); Ferrusquía-Villafranca 1990, map (geomorphological regionalization); Ramírez-Pulido and Castro-Campillo 1990, map (zoogeographic regionalization); Rzedowski and Reyna-Trujillo 1990, map (phytogeographic regionalization); Anderson and O'Brien 1996, 332 (biotic evolution); Ayala et al. 1996, 429 (zoogeographic regionalization); Arriaga et al. 1997, 64 (biogeographic regionalization); Escalante et al. 1998, 285 (cluster analysis and biogeographic regionalization); Campbell 1999, 114 (zoogeographic regionalization); Luna Vega et al. 1999, 1301 (parsimony analysis of endemicity); Morrone et al.

1999, 510 (parsimony analysis of endemicity); Espinosa et al. 2000, 64 (parsimony analysis of endemicity and biogeographic regionalization); Flores-Villela and Goyenechea 2001, 174 (cladistic biogeography); Morrone 2001, 34 (biogeographic regionalization); Morrone and Márquez 2001, 636 (track analysis); Morrone et al. 2002, 91 (biogeographic regionalization); Huber and Riina 2003, 260 (phytogeographic glossary); Corona and Morrone 2005, 38 (track analysis); Escalante et al. 2005, 202 (parsimony analysis of endemicity); Morrone 2005, 234 (biogeographic regionalization); Morrone and Gutiérrez 2005, 1317 (track analysis); Andrés Hernández et al. 2006, 901 (track analysis); Morrone 2006, 476 (biogeographic regionalization); Morrone and Llorente Bousquets 2006, 1021 (biogeographic regionalization); Contreras-Medina et al. 2007b, 408 (parsimony analysis of endemicity); Mariño-Pérez et al. 2007, 80 (track analysis); Espinosa Organista et al. 2008, 56 (biogeographic regionalization); Morrone and Márquez 2008, 19 (track analysis); Corona et al. 2009, 1732 (track analysis); León-Paniagua and Morrone 2009, 1942 (cladistic biogeography); Morrone 2010, 358 (biotic evolution); Domínguez-Domínguez et al. 2011, 2 (biotic evolution); Coulleri and Ferrucci 2012, 105 (track analysis); González-Elizondo et al. 2012, 354 (vegetation); Mercado-Salas et al. 2012, 459 (track analysis); Kobelkowsky-Vidrio et al. 2014, 2088 (parsimony analysis of endemicity); Morrone 2014a, 27 (biogeographic regionalization); García Díaz et al. 2015, 194 (track analysis); Cuellar-Martínez and Sosa 2016, 688 (species richness and endemicity analyses); Munguía-Lino et al. 2016, 146 (endemicity analysis); Morrone 2017a, 61 (biogeographic regionalization and biotic evolution); Morrone et al. 2017, 278 (map); García-Navarrete and Morrone 2018, 4 (parsimony analysis of endemicity); Rodríguez et al. 2018, 538 (species richness and endemicity); Morrone 2019, 36 (biogeographic regionalization).

Sierra Madre Occidental region West 1964, 368 (zoogeographic regionalization).

Sierra Madre Occidental Pine-Oak Forests ecoregion Dinerstein et al. 1995, 97 (ecological regionalization).

Madrean province (in part) Brown et al. 1998, 30 (vegetation).

Sierra Madre Occidental area Marshall and Liebherr 2000, 206 (cladistic biogeography); Katinas et al. 2004, 166 (parsimony analysis of endemicity); Flores-Villela and Martínez-Salazar 2009, 820 (cladistic biogeography).

Western Sierra Madre province Beron 2018, 630 (zoogeographic regionalization).

The Sierra Madre Occidental province (Fig. 4.6) is situated in western Mexico (states of Chihuahua, Durango, Jalisco, Nayarit, Sinaloa, Sonora, and Zacatecas), at elevations between 200 and 3000 m, with most of the area above 2000 m (Rzedowski 1978; Morrone 2006, 2014a; Kobelkowsky-Vidrio et al. 2014; Valero et al. 2019a). The Sierra Madre Occidental and the Rocky Mountains in the United States are connected by the Madrean Sky Islands, which are ca. 60 small mountains from southern Arizona and New Mexico (Kobelkowsky-Vidrio et al. 2014; Yanahan and Moore 2019).

Endemic and Characteristic Taxa Morrone (2014a) and Valero et al. (2019a) provided lists of endemic and characteristic taxa. Some examples include *Selaginella*

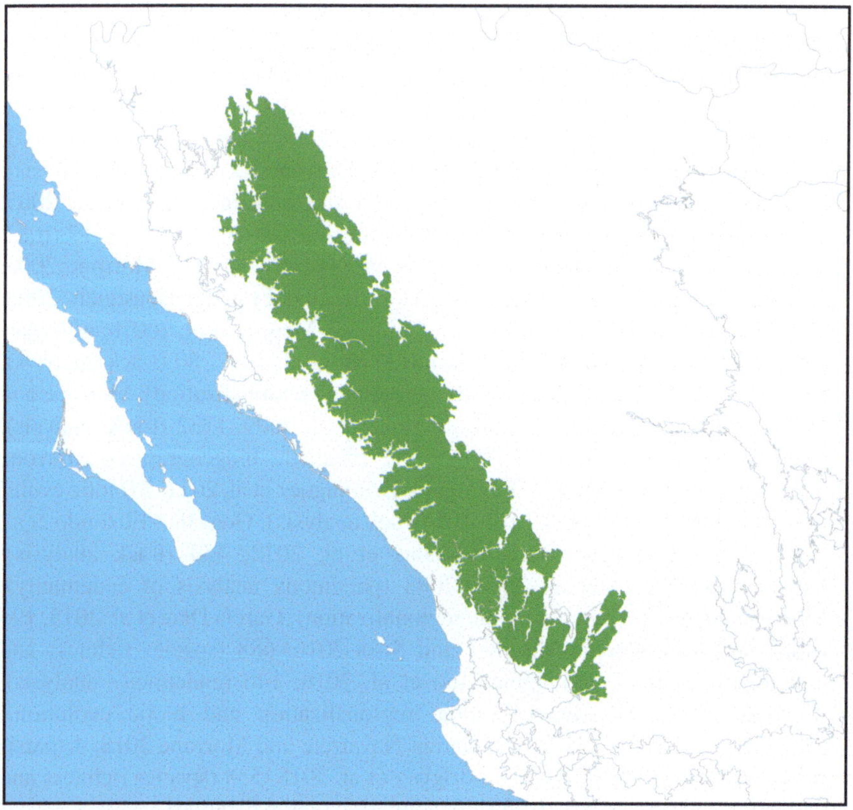

Fig. 4.6 Sierra Madre Occidental biogeographic province (Morrone et al. 2017)

mutica var. *mutica* (Selaginellaceae); *Juniperus deppeana* var. *pachyphlaea* (Cupressaceae); *Pinus engelmannii* (Pinaceae); *Quercus carmenensis*, *Q. chihuahuensis*, *Q. deliquescens*, *Q. radiata*, *Q. santaclarensis*, *Q. tarahumara*, and *Q. undata* (Fagaceae); *Elliptoleus olisthopoides* (Coleoptera: Carabidae); *Coscinocephalus cribrifrons*, *Hologymnetis argenteola*, *Homoiosternus beckeri*, *Onthophagus brevifrons*, and *Strategus cessus* (Coleoptera: Scarabaeidae); *Campostoma ornatum* (Osteichthyes: Cyprinidae); *Cyanocorax dickeyi* (Aves: Corvidae); *Rhynchopsitta pachyrhyncha* (Aves: Psittacidae); *Euptilotis neoxenus* (Aves: Trogonidae); *Peromyscus spicilegus* and *Reithrodontomys zacatecae* (Mammalia: Cricetidae); and *Glaucomys volans madrensis* and *Sciurus nayaritensis* (Mammalia: Sciuridae).

Vegetation Temperate forests, especially pine and pine-oak forests, are predominant (Rzedowski 1978; Dinerstein et al. 1995; González-Elizondo et al. 2012; Valero et al. 2019a). The montane forests of the Sierra Madre Occidental province harbor some of the world's most extensive coniferous forests (Dinerstein et al.

1995). Dominant plants include species of *Abies*, *Pinus*, and the southernmost present-day distribution of *Picea* (Graham 2004). Endemic plant genera include *Arnicastrum*, *Pionocarpus*, *Pippenalia*, *Stenocarpha*, and *Trichocoryne* (Rzedowski 1978). González-Elizondo et al. (2012) identified the following vegetation types: pine forests (1600–3320 m) with *Pinus arizonica*, *P. engelmannii*, *P. strobiformis*, *P. durangensis*, *P. teocote*, and *P. cooperi*; mixed conifer forests (1900–3300 m), dominated by species of *Abies*, *Pseudotsuga*, and *Picea*, frequently with *Pinus* and *Quercus*; pine-oak forests (1250–3200 m) with *Pinus arizonica*, *P. engelmannii*, *P. strobiformis*, *Quercus rugosa*, *Q. gambelii*, and *Q. hypoleuca*; oak forests (340–2900 m) with *Q. sideroxyla*, *Q. rugosa*, *Q. durifolia*, *Q. scytophylla*, *Q. crassifolia*, *Q. diversifolia*, *Q. albocincta*, and *Q. oblongifolia*; cloud forests (1000–2350 m) with *Magnolia pacifica*, *Styrax ramirezii*, *Cedrela odorata*, *Persea liebmannii*, *Quercus candicans*, *Q. castanea*, *Q. splendens*, and *Arbutus xalapensis*; scrub (1900–2500 m) with *Arctostaphylos pungens*, *Quercus depressipes*, and *Garrya wrightii*; and grasslands (1200–2200 m) with species of Poaceae.

Biotic Relationships According to parsimony analyses of endemicity based on plant, insect, and bird taxa (Morrone et al. 1999) and mammals (Escalante et al. 2005), the Sierra Madre Occidental is closely related to other areas in the Nearctic region. A track analysis based on species of Coleoptera (Morrone and Márquez 2001) showed that this province is related to the Balsas Basin, Sierra Madre del Sur, Sierra Madre Oriental, and Transmexican Volcanic Belt provinces. A cladistic biogeographic analysis, based on the Carabid (Coleoptera) genera *Elliptoleus* and *Calathus* (Liebherr 1991), showed a close relationship between the Sierra Madre Occidental and the northern part of the Sierra Madre Oriental.

Regionalization Smith (1941) and Moore (1945) have delimited nested units within the Sierra Madre Occidental province. I (Morrone 2014a) accepted Moore's (1945) two districts, but noting that Smith's (1941) names have nomenclatural priority.

4.3.1 Apachian District

Apachian province Smith 1941, 109 (zoogeographic regionalization).

Apachian district Goldman and Moore 1945, 353 (zoogeographic regionalization); Morrone 2014a, 29 (biogeographic regionalization); Morrone 2017a, 64 (biogeographic regionalization and biotic evolution); Morrone 2019, 38 (biogeographic regionalization).

Tarahumara district Goldman and Moore 1945, 353 (zoogeographic regionalization); Moore 1945, 218 (zoogeographic regionalization).

Sierra Madre Occidental, N province Escalante et al. 1998, 284 (cluster analysis and biogeographic regionalization).

Sierra Madre Occidental-Central Plateau area (in part) Marshall and Liebherr 2000, 205 (cladistic biogeography).

Tarahumara province Santiago-Alvarado 2019, 34 (endemicity analysis and biogeographic regionalization).

The Apachian district corresponds to the northern portion of the Sierra Madre Occidental province, in the states of Chihuahua and Sonora and part of Sinaloa. Endemic taxa include *Cheilanthes arizonica* (Pteridaceae), *Juniperus deppeana* var. *pachycephala* and *J. scopulorum* (Cupressaceae), and *Quercus tarahumara* and *Q. toumeyi* (Fagaceae).

4.3.2 Durangoan District

Durangoan province Smith 1941, 109 (zoogeographic regionalization).

Tepehuane district Goldman and Moore 1945, 353 (zoogeographic regionalization); Moore 1945, 218 (zoogeographic regionalization).

Sierra Madre Occidental, S province Escalante et al. 1998, 284 (cluster analysis and biogeographic regionalization).

Sierra Madre Occidental-Central Plateau area (in part) Marshall and Liebherr 2000, 205 (cladistic biogeography).

Durangoan district Morrone 2014a, 29 (biogeographic regionalization); Morrone 2017a, 64 (biogeographic regionalization and biotic evolution); García-Navarrete and Morrone 2018, 10 (parsimony analysis of endemicity); Morrone, 2019, 38 (biogeographic regionalization).

The Durangoan district corresponds to the southern portion of the Sierra Madre Occidental province, in the states of Jalisco, Nayarit, Sinaloa, Sonora, and Zacatecas and parts of Sinaloa, Durango, and Chihuahua. Endemic taxa include *Quercus radiata* and *Q. undata* (Fagaceae).

Kobelkowsky-Vidrio et al. (2014) undertook a parsimony analysis of endemicity of bird species, analyzing both migrant and resident species, with grid cells of 0.38° and 0.53°. The cladograms obtained (Fig. 4.7) resulted mostly unresolved, with few clades supported by endemic species. Based on these results, the authors concluded that there is a separation along the highest elevations of the Sierra Madre Occidental between a southwestern area in the Pacific slope and a northeastern area in the eastern slope. They found that this boundary coincides with the limit between the tropical and temperate vegetation and that bird species from the Pacific slope have Neotropical affinities, whereas those from the eastern slope have Nearctic affinities. They concluded that the area situated toward the Pacific slope could be regarded as part of the Mexican Transition Zone, whereas the area including the eastern slope corresponds to the Nearctic region. These results are quite interesting, but they conflict with the already recognized districts. More taxa should be analyzed to solve this issue.

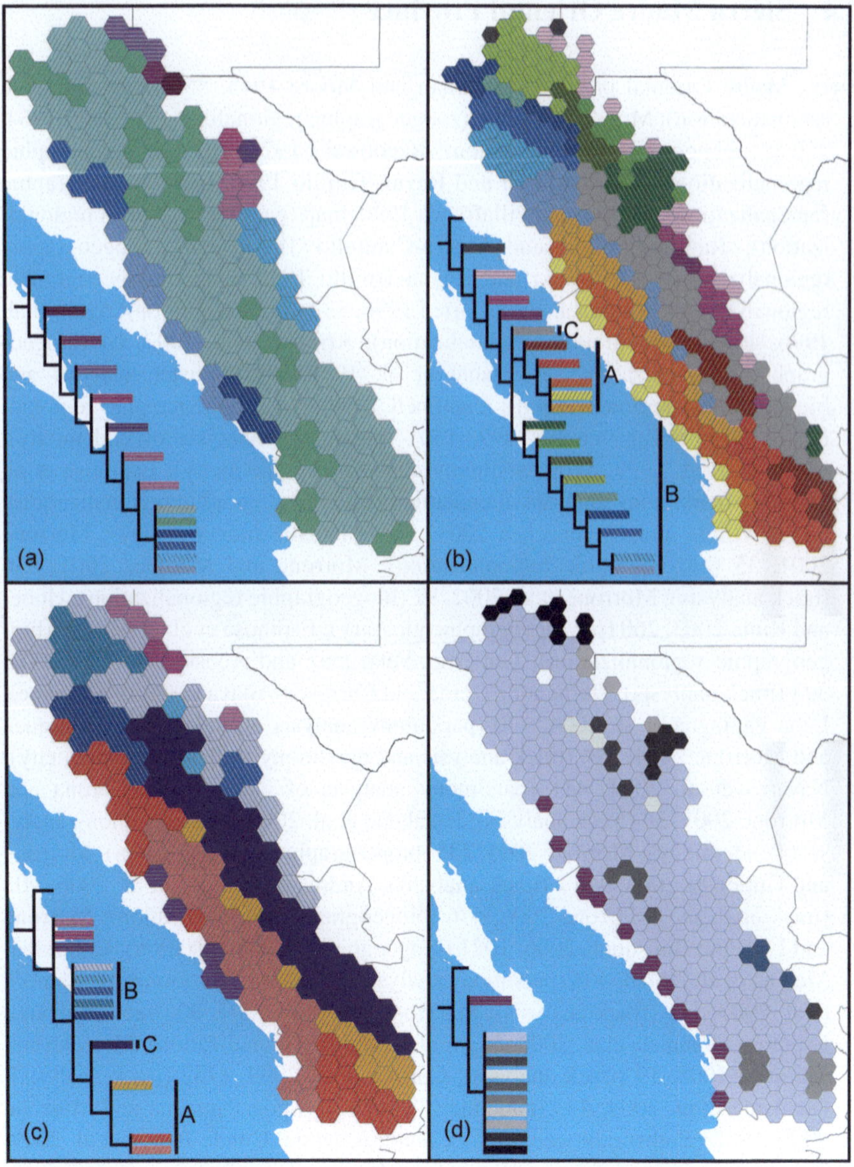

Fig. 4.7 Parsimony analysis of endemicity of the Sierra Madre Oriental (modified from Kobelkowsky-Vidrio et al. 2014). (**a**) Resident terrestrial bird species in a 0.53° grid cell; (**b**) resident terrestrial bird species in a 0.38° grid cell; (**c**) bird species with pine-oak preference in a 0.53° grid cell; (**d**) bird species with pine-oak preference in a 0.38° grid cell

4.4 Sierra Madre Oriental Province

Sierra Madre Oriental province Goldman and Moore 1945, 356 (zoogeographic regionalization); Moore 1945, 218 (zoogeographic regionalization); Stuart 1964, 350 (zoogeographic regionalization); Rzedowski 1978, 103 (phytogeographic regionalization); Casas-Andreu and Reyna-Trujillo 1990, map (zoogeographic regionalization); Ferrusquía-Villafranca 1990, map (geomorphological regionalization); Ramírez-Pulido and Castro-Campillo 1990, map (zoogeographic regionalization); Rzedowski and Reyna-Trujillo 1990, map (phytogeographic regionalization); Anderson and O'Brien 1996, 332 (biotic evolution); Ayala et al. 1996, 429 (zoogeographic regionalization); Arriaga et al. 1997, 64 (biogeographic regionalization); Escalante et al. 1998, 285 (cluster analysis and biogeographic regionalization); Campbell 1999, 114 (zoogeographic regionalization); Luna Vega et al. 1999, 1301 (parsimony analysis of endemicity); Morrone et al. 1999, 510 (parsimony analysis of endemicity); Espinosa et al. 2000, 64 (parsimony analysis of endemicity and biogeographic regionalization); Flores-Villela and Goyenechea 2001, 174 (cladistic biogeography); Morrone 2001, 35 (biogeographic regionalization); Morrone and Márquez 2001, 636 (track analysis); Morrone et al. 2002, 92 (biogeographic regionalization); Huber and Riina 2003, 260 (phytogeographic glossary); Espinosa et al. 2004, 487 (biogeographic regionalization); Gutiérrez-Velázquez and Acosta-Gutiérrez 2004, 393 (track analysis); Hernández-Cerda and Carrasco-Anaya 2004, 63 (climate); León-Paniagua et al. 2004, 470 (parsimony analysis of endemicity); Márquez and Morrone 2004, 376 (track analysis and parsimony analysis of endemicity); Navarro et al. 2004, 441 (parsimony analysis of endemicity); Corona and Morrone 2005, 38 (track analysis); Escalante et al. 2005, 202 (parsimony analysis of endemicity); Morrone 2005, 235 (biogeographic regionalization); Morrone and Gutiérrez 2005, 1317 (track analysis); Andrés Hernández et al. 2006, 902 (track analysis); Morrone 2006, 476 (biogeographic regionalization); Morrone and Llorente Bousquets 2006, 1021 (biogeographic regionalization); Contreras-Medina et al. 2007b, 408 (parsimony analysis of endemicity); González-Zamora et al. 2007, 136 (track analysis); Mariño-Pérez et al. 2007, 80 (track analysis); Espinosa Organista et al. 2008, 56 (biogeographic regionalization); Morrone and Márquez 2008, 19 (track analysis); Corona et al. 2009, 1732 (track analysis); Escalante et al. 2009, 473 (endemicity analysis); León-Paniagua and Morrone 2009, 1942 (cladistic biogeography); Santa Anna del Conde Juárez et al. 2009, 374 (parsimony analysis of endemicity); Morrone 2010, 358 (biotic evolution); Sanginés-Franco et al. 2011, 82 (parsimony analysis of endemicity); Coulleri and Ferrucci 2012, 105 (track analysis); Puga-Jiménez et al. 2013, 1180 (parsimony analysis of endemicity); Morrone 2014a, 29 (biogeographic regionalization); García Díaz et al. 2015, 192 (track analysis); Cuellar-Martínez and Sosa 2016, 688 (species richness and endemicity analyses); Ferro et al. 2017, 2146 (biotic evolution); Morrone 2017a, 65 (biogeographic regionalization and biotic evolution); Morrone et al. 2017, 278 (map); Salinas-Rodríguez et al. 2017, 1 (vegetation); García-Navarrete and Morrone 2018, 4 (parsimony analysis of endemicity); Rodríguez et al. 2018, 538 (species richness and endemicity); de la Rosa-Manzano et al. 2019, 2 (vegetation); Estrada Sánchez et al. 2019, 133 (endemicity analysis); Morrone 2019, 38 (biogeographic regionalization); Santiago-Alvarado 2019, 34 (endemicity analysis and biogeographic regionalization).

Sierra Madre Oriental region West 1964, 368 (zoogeographic regionalization).

Sierra Madre Oriental Pine-Oak Forests ecoregion Dinerstein et al. 1995, 97 (ecological regionalization).

Madrean province (in part) Brown et al. 1998, 30 (vegetation).

Sierra Madre Oriental area Marshall and Liebherr 2000, 206 (cladistic biogeography); Katinas et al. 2004, 166 (parsimony analysis of endemicity); Flores-Villela and Martínez-Salazar 2009, 820 (cladistic biogeography).

Eastern Sierra Madre province (Beron 2018, 630 (zoogeographic regionalization).

The Sierra Madre Oriental province (Fig. 4.8) is situated in eastern Mexico (states of Coahuila, Durango, Guanajuato, Hidalgo, Nuevo León, Puebla, Querétaro, San Luis Potosí, Tamaulipas, Veracruz, and Zacatecas), at elevations above 1500 m (Rzedowski 1978; Morrone 2014a; Luna et al. 2004; Sanginés-Franco et al. 2011; Valero et al. 2019b). The Sierra Madre Oriental physiographic province is the second largest montane system in Mexico, with 1350 km length and average altitudes of 1500–2000 m, in some places reaching 3000 m. The tallest peaks are Cerro Potosí (3625 m) and Peña Nevada (3480 m) (Valero et al. 2019b). It runs northwest to southeast, beginning in the Sierra del Burro, close to the Bravo River in Tamaulipas and ending in the Cofre de Perote (Veracruz), where it connects with the Transmexican Volcanic Belt (González-Zamora et al. 2007). The windward slope is wet and influenced by marine winds, and the leeward slope has a drier climate (de la Rosa-Manzano et al. 2019).

Endemic and Characteristic Taxa Morrone (2014a) and Valero et al. (2019b) provided lists of endemic and characteristic taxa. Some examples include *Cheilanthes decomposita* (Pteridaceae); *Pinus arizonica*, *P. cembroides*, *P. greggii*, *P. nelsonii*, and *P. patula* (Pinaceae); *Agave inaequidens*, *A. horrida*, and *A. tenuifolia* (Asparagaceae); *Priamides erostratus erostratinus* and *Pterourus palamedes leontis* (Lepidoptera: Papilionidae); *Odontotaenius zodiacus*, *Petrejoides laticornis*, *P. nebulosus*, *P. orizabae*, and *P. silvaticus* (Coleoptera: Passalidae); *Typhlochactas* spp. (Scorpiones: Superstitionidae); *Glaucidium sanchezi* (Aves: Strigidae); *Peromyscus aztecus* and *Reithrodontomys fulvescens* (Mammalia: Cricetidae); and *Cryptotis obscura* and *C. parva berlandieri* (Mammalia: Soricidae). Salinas-Rodríguez et al. (2017) provided a list of 1135 endemic vascular plants, being the most specious families Asteraceae (232 species), Cactaceae (170), Fabaceae (66), and Lamiaceae (66).

Vegetation It consists of temperate forests (Fig. 4.9), with pine-oak forests being predominant (Valero et al. 2019b). There are also semideciduous and cloud forests (Dinerstein et al. 1995; Sanginés-Franco et al. 2011), as well as areas with xero-

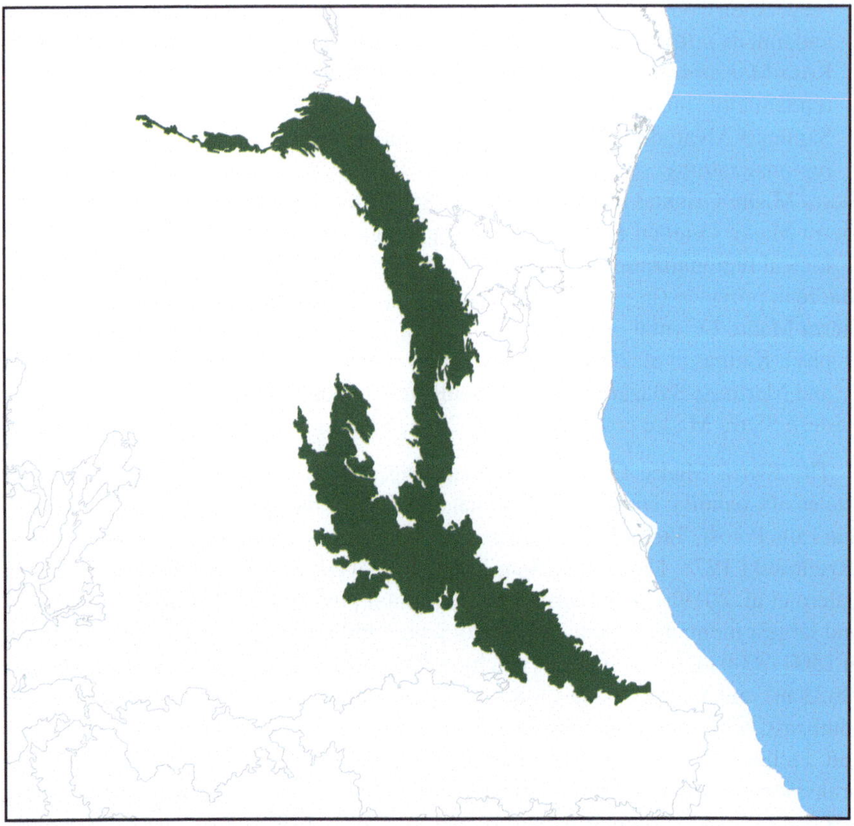

Fig. 4.8 Sierra Madre Oriental biogeographic province (Morrone et al. 2017)

phytic scrubland, mainly in the lowlands near the Mexican Plateau (leeward slope), where cacti are well represented (Santa Anna del Conde Juárez et al. 2009). De la Rosa-Manzano et al. (2019) characterized four vegetation types along a gradient from 300 to 2000 m: semideciduous forest (300–400 m) with a warm and subhumid climate, dominated by *Bursera simaruba*, *Casimiroa greggii*, *Lysiloma divaricatum*, *Cascabela thevetia*, and *Trema micrantha*; tropical montane cloud forest (800–1600 m) with a warmer climate, dominated by *Magnolia tamaulipana*, *Quercus germana*, *Podocarpus matudae*, *Acer skutchii*, and *Carpinus caroliniana*; submontane scrub (700–1500 m) with a dry climate, dominated by *Acacia berlandieri*, *Acanthocereus tetragonus*, *Cordia boissieri*, *Gochnatia hypoleuca*, and *Yucca treculeana*; and pine-oak forest (1500–2000 m) with a dry and cool climate, dominated by *Pinus teocote*, *P. pseudostrobus*, *Carya ovata*, *Quercus germana*, *Q. affinis*, and *Q. sartorii*. Semideciduous and tropical montane cloud forests are distributed along the windward slope of the Sierra Madre Oriental, whereas the submontane scrub and pine-oak forests are distributed along the leeward slope.

Fig. 4.9 Vegetation of the Sierra Madre Oriental province. (**a**) Pine forest, Miquihuana, Tamaulipas; (**b**) cloud forest, Tenango de Doria, Hidalgo (pictures by Juan Márquez)

Biotic Relationships According to a parsimony analysis of endemicity based on plant, insect, and bird taxa (Morrone et al. 1999), the Sierra Madre Oriental province is closely related to the southern portion of the Mexican Plateau province. Another parsimony analysis of endemicity based on mammals (Escalante et al. 2005) found that it is related to the Transmexican Volcanic Belt province. A track analysis based on species of Coleoptera (Morrone and Márquez 2001) showed that this province is related to the Sierra Madre Occidental, Sierra Madre del Sur, Balsas Basin, and Transmexican Volcanic Belt provinces. Another track analysis (Sanginés-Franco et al. 2011), based on fern species, highlighted the complex biotic relationships of the Saltillo-Parras district. A parsimony analysis of endemicity based on bird taxa (Sánchez-González et al. 2008), a cladistic biogeographic analysis based on two Carabid genera (Liebherr 1991), and the phylogeographic analysis of the furnariid bird *Lepidocolaptes affinis* (Arbeláez-Cortés et al. 2010) have suggested that the Sierra Madre Oriental province may be a composite of two portions. The northern one is related to the Sierra Madre Occidental, and the southern one is related to the Sierra Madre del Sur and the Transmexican Volcanic Belt.

Ferro et al. (2017) analyzed the bird species distributed in the Sierra Madre Oriental province, in order to detect their biogeographic affinities. The authors considered two different features: the chorological affinity based on raw distributional data concordance (chorotypes) and the evolutionary biogeographic affinity based on the ancestral state reconstruction on published phylogenetic analyses (cenocrons). The authors identified the sharpest transition at elevations of 1500 m from the southern cloud forests to the canyons of the middle Pánuco basin, then turning eastward, following the Pánuco river through the semideciduous lowland forest to the river mouth in the Gulf of Mexico. This confirmed that the middle elevation areas of the mountains of the Mexican Transition Zone host a mixture of Nearctic and Neotropical biotas.

Regionalization Espinosa et al. (2004), Espinosa Organista et al. (2008), and Morrone (2014a, 2017a, 2019) have recognized two subprovinces and four districts within the Sierra Madre Oriental province. More recently, Santiago-Alvarado (2019) recognized smaller districts, which I consider provisionally as part of the previously recognized.

4.4.1 Austral-Oriental Subprovince

Austral-Oriental province Smith 1941, 108 (zoogeographic regionalization).
Sierra Madre Oriental N province Escalante et al. 1998, 284 (cluster analysis and biogeographic regionalization).
Meridional subprovince Espinosa et al. 2004, 294 (biogeographic regionalization).
Austral-Oriental subprovince Morrone 2014a, 30 (biogeographic regionalization); 2017a, 68 (biogeographic regionalization and biotic evolution); 2019, 40 (biogeographic regionalization).

The Austral-Oriental subprovince corresponds to the northern portion of the Sierra Madre Oriental province, north of the Moctezuma River (Espinosa et al. 2004). Some endemic taxa include *Ageratina potosina*, *Flourensia monticola*, *Flyriella stanfordii*, *Grindelia greenmanii*, *Porophyllum filiforme*, *Rumfordia exauriculata*, *Senecio carnerensis*, *Stevia hintoniorum*, and *Verbesina daviesiae* (Asteraceae), *Sciurus alleni* (Mammalia: Sciuridae), and *Sorex milleri* (Mammalia: Soricidae) (Espinosa et al. 2004; Ceballos and Oliva 2005; González-Zamora et al. 2007). Within this subprovince, two districts have been recognized.

4.4.1.1 Saltillo-Parras District

Saltillo-Parras district Espinosa et al. 2004, 294 (biogeographic regionalization); Morrone 2014a, 31 (biogeographic regionalization); 2017a, 69 (biogeographic regionalization and biotic evolution); 2019, 40 (biogeographic regionalization).

Sierras Transversales district Espinosa Organista et al. 2008, 56 (biogeographic regionalization); Santiago-Alvarado 2019, 34 (endemicity analysis and biogeographic regionalization).

Sierras Transversales Gran Sierra Plegada Norte district Santiago-Alvarado 2019, 34 (endemicity analysis and biogeographic regionalization).

The Saltillo-Parras district corresponds to the Pliegues Saltillo-Parras, Sierras Transversales, and the northern part of the Gran Sierra Plegada (Espinosa et al. 2004). Some endemic taxa include *Pinus culminicola* (Pinaceae); *Acharagma roseana*, *Mammillaria grusonii*, and *Turbinicarpus booleanus* (Cactaceae); *Quercus sinuata breviloba* (Fagaceae); *Pseudoeurycea galeana* (Amphibia: Plethodontidae); *Peromyscus hooperi* (Mammalia: Cricetidae); and *Myotis planiceps* (Mammalia: Vespertilionidae) (Canseco-Márquez et al. 2004; Espinosa et al. 2004; Ceballos and Oliva 2005; Santa Anna del Conde Juárez et al. 2009).

4.4.1.2 Potosí District

Potosí district Espinosa et al. 2004, 294 (biogeographic regionalization); Morrone 2014a, 30 (biogeographic regionalization); 2017a, 69 (biogeographic regionalization and biotic evolution); 2019, 40 (biogeographic regionalization).

Gran Sierra Plegada district Espinosa Organista et al. 2008, 56 (biogeographic regionalization); Santiago-Alvarado 2019, 35 (endemicity analysis and biogeographic regionalization).

Sierra de Cuchar district Santiago-Alvarado 2019, 35 (endemicity analysis and biogeographic regionalization).

The Potosí district corresponds to the Gran Sierra Plegada and parts of the Sierras and Llanuras Occidentales, at Sierra de Potosí (Espinosa et al. 2004). Some endemic taxa include *Phanerophlebia umbonata* (Dryopteridaceae); *Dioon angustifolium* (Zamiaceae); *Pinus nelsonii* (Pinaceae); *Quercus canbyi* and *Q. rysophylla* (Fagaceae); *Guazuma ulmifolia* (Malvaceae); *Chiropterotriton cracens*, *C. multidentatus*, and *Pseudoeurycea scandens* (Amphibia: Plethodontidae); *Neotoma angustapalata* and *Peromyscus ochraventer* (Mammalia: Cricetidae), and *Notiosorex villai* (Mammalia: Soricidae) (Canseco-Márquez et al. 2004; Espinosa et al. 2004; Ceballos and Oliva 2005; Santiago-Alvarado 2019).

4.4.2 *Hidalgoan Subprovince*

Hidalgoan province Smith 1941, 108 (zoogeographic regionalization).
Sierra Madre Oriental S province Escalante et al. 1998, 284 (cluster analysis and biogeographic regionalization).
Septentrional subprovince Espinosa et al. 2004, 294 (biogeographic regionalization).
Hidalgoan subprovince Morrone 2014a, 30 (biogeographic regionalization); 2017a, 69 (biogeographic regionalization and biotic evolution); 2019, 40 (biogeographic regionalization).
Carso Huasteco-Sierra de Juárez province Santiago-Alvarado 2019, 35 (endemicity analysis and biogeographic regionalization).

The Hidalgoan subprovince corresponds to the southern portion of the Sierra Madre Oriental province, south of the Moctezuma River, in the Carso Huasteco physiographic province (Espinosa et al. 2004). Some endemic taxa include *Archibaccharis venturana*, *Chaptalia estribensis*, and *Verbesina coulteri* (Asteraceae), *Inga huastecana* (Fabaceae), *Prunus samydoides* (Rosaceae), *Dendrortyx barbatus* (Aves: Odontophoridae), and *Megadontomys nelsoni*, *Microtus quasiater*, *Peromyscus furvus*, and *Reithrodontomys sumichrasti* (Mammalia: Cricetidae) (Espinosa et al. 2004; Ceballos and Oliva 2005; Ramírez-Pulido et al. 2005; González-Zamora et al. 2007; Santiago-Alvarado 2019). Within this subprovince, two districts have been recognized.

4.4.2.1 Sierra Gorda District

Sierra Gorda district Espinosa et al. 2004, 294 (biogeographic regionalization); Morrone 2014a, 30 (biogeographic regionalization); 2017a, 69 (biogeographic regionalization and biotic evolution); 2019, 40 (biogeographic regionalization).

The Sierra Gorda district corresponds to the western part of the subprovince, delimited by the rivers Moctezuma and Verde, which are tributaries of the Pánuco River (Espinosa et al. 2004). Some endemic taxa include *Agave tenuifolia* (Asparagaceae), *Styrax argenteus* subsp. *parvifolius* (Styracaceae), *Sciurus oculatus oculatus* (Mammalia: Sciuridae), and *Euderma phyllote phyllote* (Mammalia: Vespertilionidae) (Espinosa et al. 2004).

4.4.2.2 Zacualtipán District

Zacualtipán district Espinosa et al. 2004, 294 (biogeographic regionalization); Morrone 2014a, 30 (biogeographic regionalization); 2017a, 69 (biogeographic regionalization and biotic evolution); 2019, 40 (biogeographic regionalization).

Carso Huasteco Sur-Sierra de Juárez district Santiago-Alvarado 2019, 36 (endemicity analysis and biogeographic regionalization).
Sierra de Chiconquiaco Norte district Santiago-Alvarado 2019, 36 (endemicity analysis and biogeographic regionalization).
Sierra de Chiconquiaco-Zongolica district Santiago-Alvarado 2019, 36 (endemicity analysis and biogeographic regionalization).
Sierra Nevada-Chiconquiaco-Norte Sierra de Juárez nodal province Santiago-Alvarado 2019, 42 (endemicity analysis and biogeographic regionalization).

The Zacualtipán district corresponds to the eastern part of the subprovince, including the Sierra Norte de Puebla and the Sierra de Zacualtipán (Espinosa et al. 2004). Some endemic taxa include *Campyloneurum angustifolium* (Polypodiaceae); *Pinus pseudostrobus apulcensis* (Pinaceae); *Fleischmannia pycnocephala* (Asteraceae); *Quercus hirtifolia* (Fagaceae); *Aporocactus flagelliformis* (Cactaceae); *Smilax mollis* (Smilacaceae); *Pseudoeurycea cafetalera*, *P. lynchi*, and *P. quetzalanensis* (Amphibia: Plethodontidae); and *Habromys simulatus* (Mammalia: Cricetidae) (Espinosa et al. 2004)

4.5 Transmexican Volcanic Belt Province

Austral-Western province Smith 1941, 108 (zoogeographic regionalization); Huber and Riina 2003, 257 (phytogeographic glossary).
Transverse Volcanic province Goldman and Moore 1945, 356 (zoogeographic regionalization); Moore 1945, 218 (zoogeographic regionalization); Stuart 1964, 351 (zoogeographic regionalization); Huber and Riina 2003, 331 (phytogeographic glossary).
Mesa Central area (in part) West 1964, 368 (zoogeographic regionalization).
Meridional Mountain province (in part) Rzedowski 1978, 103 (phytogeographic regionalization); Rzedowski and Reyna-Trujillo 1990, map (phytogeographic regionalization); Luna Vega et al. 1999, 1301 (parsimony analysis of endemicity); Huber and Riina 2003, 260 (phytogeographic glossary).
Neovolcanic Axis province Casas-Andreu and Reyna-Trujillo 1990, map (zoogeographic regionalization); Escalante et al. 1998, 285 (cluster analysis and biogeographic regionalization); Espinosa Organista et al. 2008, 56 (biogeographic regionalization); García Díaz et al. 2015, 192 (track analysis).
Neovolcanic province Ferrusquía-Villafranca 1990, map (geomorphological regionalization).
Mexican Alpine Tundra ecoregion Dinerstein et al. 1995, 101 (ecological regionalization).
Mexican Transvolcanic Pine-Oak Forests ecoregion Dinerstein et al. 1995, 97 (ecological regionalization).
Transverse Volcanic Belt province Anderson and O'Brien 1996, 332 (biotic evolution); Ayala et al. 1996, 429 (zoogeographic regionalization).

Volcanic Axis province Arriaga et al. 1997, 64 (biogeographic regionalization); Morrone et al. 1999, 510 (parsimony analysis of endemicity); Espinosa et al. 2000, 64 (parsimony analysis of endemicity and biogeographic regionalization).

Transvolcanic province Brown et al. 1998, 29 (vegetation).

Sierra Transvolcánica area Marshall and Liebherr 2000, 206 (cladistic biogeography).

Transvolcanic Axis area Flores-Villela and Goyenechea 2001, 174 (cladistic biogeography).

Transmexican Volcanic Belt province Morrone 2001, 35 (biogeographic regionalization); Morrone and Márquez 2001, 636 (track analysis); Morrone et al. 2002, 93 (biogeographic regionalization); Corona and Morrone 2005, 38 (track analysis); Escalante et al. 2005, 202 (parsimony analysis of endemicity); Morrone 2005, 236 (biogeographic regionalization); Morrone and Gutiérrez 2005, 1317 (track analysis); Andrés Hernández et al. 2006, 902 (track analysis); Morrone 2006, 477 (biogeographic regionalization); Morrone and Llorente 2006, 1021 (biogeographic regionalization); Torres Miranda and Luna Vega 2006, 849 (track analysis and biogeographic regionalization); Corona et al. 2007, 1008 (track analysis); Escalante et al. 2007b, 486 (biogeographic regionalization); Espinosa and Ocegueda 2007, 6 (biogeographic regionalization); Ferrusquía-Villafranca 2007, 8 (paleogeography); García Calderón et al. 2007, 73 (soils); Hernández-Cerda and Carrasco-Anaya 2004, 57 (climate); Mariño-Pérez et al. 2007, 80 (track analysis); Martínez-Aquino et al. 2007, 449 (track analysis); Navarro-Sigüenza et al. 2007, 462 (parsimony analysis of endemicity); Velasco de León et al. 2007, 25 (geology); Morrone and Márquez 2008, 19 (track analysis); Corona et al. 2009, 1732 (track analysis); Morrone 2010, 358 (biotic evolution); Barrera-Moreno et al. 2011, 15 (track analysis); Coulleri and Ferrucci 2012, 105 (track analysis); Gámez et al. 2012, 259 (biogeographic regionalization); Mercado-Salas et al. 2012, 459 (track analysis); Puga-Jiménez et al. 2013, 1180 (parsimony analysis of endemicity); Suárez-Mota et al. 2013, 94 (cluster analysis and biogeographic regionalization); Morrone 2014a, 31 (biogeographic regionalization); Cuellar-Martínez and Sosa 2016, 688 (species richness and endemicity analyses); Munguía-Lino et al. 2016, 146 (endemicity analysis); Morrone 2017a, 69 (biogeographic regionalization and biotic evolution); Morrone et al. 2017, 278 (map); Arriaga-Jiménez et al. 2018, 3 (species richness and diversity); Beron 2018, 630 (zoogeographic regionalization); García-Navarrete and Morrone 2018, 4 (parsimony analysis of endemicity); Rodríguez et al. 2018, 538 (species richness and endemicity); Estrada Sánchez et al. 2019, 133 (endemicity analysis); Morrone 2019, 42 (biogeographic regionalization).

Transverse Volcanic Axis province Huber and Riina 2003, 262 (phytogeographic glossary).

Trans-Mexican Volcanic area Katinas et al. 2004, 166 (parsimony analysis of endemicity).

Transvolcanic area Flores-Villela and Martínez-Salazar 2009, 820 (cladistic biogeography).

The Transmexican Volcanic Belt province (Fig. 4.10) is situated in central Mexico (states of Aguascalientes, Mexico City, Guanajuato, Jalisco, Mexico, Michoacán, Oaxaca, Puebla, Tlaxcala, and Veracruz), at elevations above 1800 m (Morrone 2006, 2014a; Luna et al. 2007; Gámez et al. 2012; Valero et al. 2019c). It corresponds to a large Neogene volcanic arc, encompassing 160,000 km^2 and a length of almost 1000 km, situated between 18°30′ and 21°30'N. The Transmexican Volcanic Belt was built upon Cretaceous and Cenozoic magmatic provinces and a heterogeneous basement made of tectonostratigraphic terranes of different age and lithology (Ferrari et al. 2012). It originated during the Miocene around central Mexico and then extended eastward and westward (Ruiz-Sánchez and Specht 2013).

Endemic and Characteristic Taxa Espinosa Organista et al. (2008), Morrone (2014a), and Valero et al. (2019c) provided lists of endemic and characteristic taxa. Some examples include *Elaphoglossum rufescens* (Dryopteridaceae); *Pinus ayacahuite* var. *veitchii*, *P. montezumae*, *P. oocarpa*, and *P. rudis* (Pinaceae); *Agave horrida* (Asparagaceae); *Montanoa frutescens* (Asteraceae); *Quercus conspersa*, *Q. flaccida*, *Q. magnifolia*, and *Q. peduncularis* (Fagaceae); *Elliptoleus balli*, *E. corvus*, and *E. vixstriatus* (Coleoptera: Carabidae); *Odontotaenius cuspidatus*

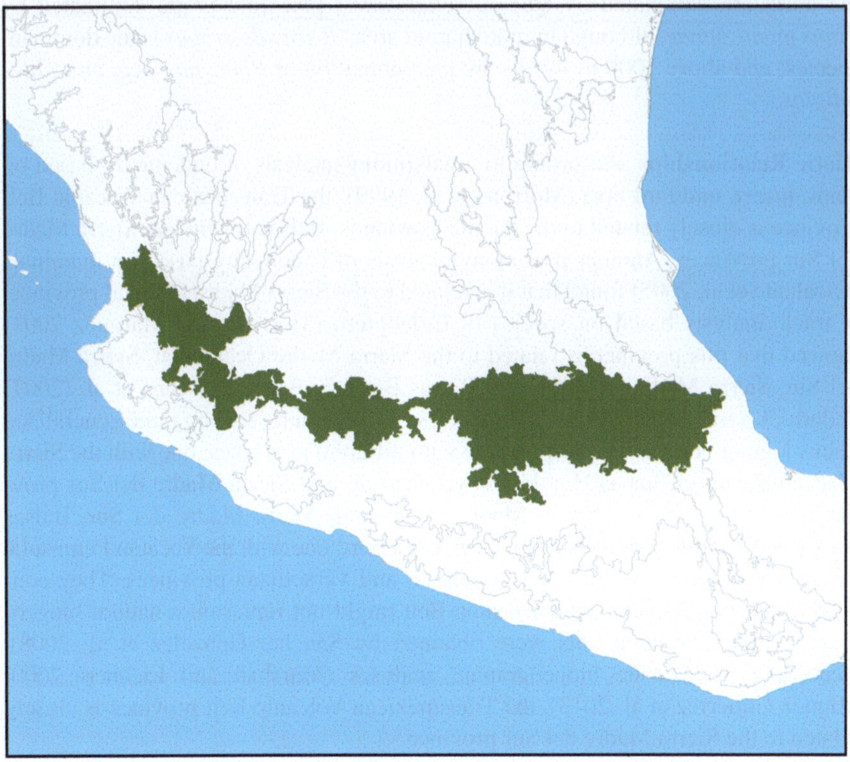

Fig. 4.10 Transmexican Volcanic Belt biogeographic province (Morrone et al. 2017)

(Coleoptera: Passalidae); *Golofa globulicornis*, *Onthophagus canescens*, and the majority of the species of the *O. hippopotamus* species group (Coleoptera: Scarabaeidae); *Quiscalus palustris* (Aves: Icteridae); *Geothlypis speciosa* and *G. trichas chapalensis* (Aves: Parulidae); *Coturnicops noveboracensis goldmani* (Aves: Rallidae); *Nelsonia goldmani*, *Peromyscus gratus*, *Reithrodontomys chrysopsis*, and *R. zacatecae* (Mammalia: Cricetidae); *Cratogeomys tylorhinus* (Mammalia: Geomyidae); and *Cryptotis alticola* and *Sorex orizabae* (Mammalia: Soricidae). Alcántara and Paniagua (2007) concluded that the Transmexican Volcanic Belt province has a total of 63 endemic species of vascular plants.

Vegetation Pine-oak (Fig. 4.11a) and oak forests are predominant (Espinosa Organista et al. 2008; Meave et al. 2016; Valero et al. 2019c). There are also cloud forests, grasslands (Fig. 4.11b), xeric shrublands, and *zacatonales* (alpine tundra) near the top of the volcanoes (Dinerstein et al. 1995). The predominant plant genera are *Achaenipodium*, *Hintonella*, *Microspermum*, *Omiltemia*, *Peyritschia*, and *Silviella* (Rzedowski 1978). Valero et al. (2019c) characterized some of the vegetation types found in this province: oak forests are more common in the western portion of the Transmexican Volcanic Belt, with *Quercus resinosa* and *Juniperus flaccida*; transition areas between pine-oak forests and deciduous forests found in the north are dominated by *Quercus glaucoides*; pine forests are dominated by *Pinus montezumae*, although in more humid areas *P. pseudostrobus* is the dominant species; and above 3000 m forests are a combination of *Pinus hartwegi* and *Abies religiosa*.

Biotic Relationships According to a parsimony analysis of endemicity based on plant, insect, and bird taxa (Morrone et al. 1999), the Transmexican Volcanic Belt province is closely related to the Pacific Lowlands, Balsas Basin, and Sierra Madre del Sur provinces. Another parsimony analysis of endemicity based on mammals (Escalante et al. 2005) found that it is related to the Sierra Madre Oriental province. A track analysis based on species of Coleoptera (Morrone and Márquez 2001) showed that this province is related to the Sierra Madre Occidental, Sierra Madre del Sur, Sierra Madre Oriental, and Balsas Basin provinces. Corona et al. (2007) undertook a track analysis based on species of Coleoptera, finding three generalized tracks joining portions of this province with different provinces: one with the Sierra Madre Oriental, Chiapas Highlands, Veracruzan, and Sierra Madre del Sur provinces; another with the Sierra Madre Occidental, Sierra Madre del Sur, Balsas Basin, and Pacific Lowlands provinces; and a third one with the Yucatán Peninsula, Chiapas Highlands, Sierra Madre Oriental, and Veracruzan provinces. They concluded that the Transmexican Volcanic Belt might not represent a natural biogeographic unit (similar results were obtained by Sánchez-González et al. 2008). According to cladistic biogeographic analyses (Marshall and Liebherr 2000; Miguez-Gutiérrez et al. 2013), the Transmexican Volcanic Belt province is closely related to the Sierra Madre del Sur province.

Fig. 4.11 Vegetation of the Transmexican Volcanic Belt province, Cerro El Pinal, Puebla. (**a**) Pine-oak forest; (**b**) grassland (pictures by Víctor Moctezuma)

Regionalization Some units nested within this province were identified by different authors (Moore 1945; Rzedowski 1978; Ferrusquía-Villafranca 1990; Escalante et al. 2007b; Navarro-Sigüenza et al. 2007; Torres Miranda and Luna 2007; Gámez et al. 2012; Suárez-Mota et al. 2013). They were treated by Morrone (2014a) as two subprovinces and five districts. As a preliminary delimitation, I (Morrone 2014a, 2017a, 2019) considered Torres Miranda and Luna's (2007) districts, but noting that

Moore's (1945) names have nomenclatural priority. One of these districts, the Jaliscian district, was later transferred to the Sierra Madre del Sur province (Morrone 2017b). More recently, Santiago-Alvarado (2019) recognized smaller districts, which I consider provisionally as part of the previous ones.

4.5.1 West Subprovince

Meridional subprovince (in part) Ferrusquía-Villafranca 1990, map (geomorphological regionalization).

Septentrional subprovince (in part) Ferrusquía-Villafranca 1990, map (geomorphological regionalization).

West district Escalante et al. 2007b, 496 (biogeographic regionalization).

West zone Suárez-Mota et al. 2013, 101 (phytogeographic regionalization).

West subprovince Morrone 2014a, 32 (biogeographic regionalization); 2017a, 72 (biogeographic regionalization and biotic evolution); 2019, 43 (biogeographic regionalization).

The West subprovince corresponds to the western portion of the Transmexican Volcanic Belt province. Endemic taxa include *Reithrodontomys goldmani* and *R. hirsutus* (Mammalia: Cricetidae), *Cratogeomys gymnurus* (Mammalia: Geomyidae), and *Liomys spectabilis* (Mammalia: Heteromyidae) (Gámez et al. 2012). Within this subprovince, two districts have been recognized.

4.5.1.1 Otomí District

Otomí district Moore 1945, 218 (zoogeographic regionalization); Morrone 2014a, 33 (biogeographic regionalization); 2017a, 72 (biogeographic regionalization and biotic evolution); García-Navarrete and Morrone 2018, 10 (parsimony analysis of endemicity); Morrone 2019; 43 (biogeographic regionalization).

The Otomí district corresponds to the central northern part of the West subprovince, in the states of Jalisco, Guanajuato, Querétaro, Hidalgo, Michoacán, and Mexico (Moore 1945). Endemic taxa include *Habromys delicatulus* (Mammalia: Cricetidae) and *Zygogeomys trichopus* (Mammalia: Geomyidae) (Ceballos and Oliva 2005; Gámez et al. 2012).

4.5.1.2 Tarascan District

Tarascan district Moore 1945, 218 (zoogeographic regionalization); Morrone 2014a, 33 (biogeographic regionalization); 2017a. 72 (biogeographic regionalization and biotic evolution); 2019, 43 (biogeographic regionalization).

The Tarascan district corresponds to the central southern part of the West subprovince, in the states of Jalisco, Michoacán, and Mexico (Moore 1945). An endemic taxon is *Reithrodontomys sumichrasti nerterus* (Mammalia: Cricetidae) (Ramírez-Pulido et al. 2005).

4.5.2 East Subprovince

Meridional subprovince (in part) Ferrusquía-Villafranca 1990, map (geomorphological regionalization).
Septentrional subprovince (in part) Ferrusquía-Villafranca 1990, map (geomorphological regionalization).
East district Escalante et al. 2007a, b, 496 (biogeographic regionalization).
East zone Suárez-Mota et al. 2013, 101 (phytogeographic regionalization).
East subprovince Morrone 2014a, 31 (biogeographic regionalization); 2017a, 72 (biogeographic regionalization and biotic evolution); García-Navarrete and Morrone 2018, 11 (parsimony analysis of endemicity); Morrone 2019. 43 (biogeographic regionalization).
Sierra Zitácuaro-Cofre de Perote province Santiago-Alvarado 2019, 36) (endemicity analysis and biogeographic regionalization).

The East subprovince corresponds to the eastern portion of the Transmexican Volcanic Belt province. Some endemic taxa include *Neotoma nelsoni* and *Peromyscus bullatus* (Mammalia: Cricetidae), *Cratogeomys merriami* (Mammalia: Geomyidae), *Spermophilus perotensis* (Mammalia: Sciuridae), and *Sorex ventralis* (Mammalia: Soricidae) (Ceballos and Oliva 2005; Gámez et al. 2012). Within this subprovince, two districts have been recognized.

4.5.2.1 Aztec District

Aztec district Moore 1945, 218 (zoogeographic regionalization); Morrone 2014a, 32 (biogeographic regionalization); 2017a, 72 (biogeographic regionalization and biotic evolution); 2019, 43 (biogeographic regionalization).
Toluca-México-Puebla district Torres Miranda and Luna 2007, 513 (track analysis and biogeographic regionalization).
Nevado de Toluca-Cofre de Perote subprovince Santiago-Alvarado 2019, 37 (endemicity analysis and biogeographic regionalization).
La Malinche-Cofre de Perote district Santiago-Alvarado 2019, 38 (endemicity analysis and biogeographic regionalization).
Nevado de Toluca-Iztaccíhuatl-Popocatépetl district Santiago-Alvarado 2019, 37 (endemicity analysis and biogeographic regionalization).
Nevado de Toluca-Sierra de Taxco district Santiago-Alvarado 2019, 38) (endemicity analysis and biogeographic regionalization).

Nevado de Toluca-Sierra Nevada district Santiago-Alvarado 2019, 37 (endemicity analysis and biogeographic regionalization).

The Aztec district corresponds to the western part of the East subprovince, in the states of Mexico, Hidalgo, Tlaxcala, Puebla, Morelos, and Guerrero (Moore 1945). Some endemic taxa include *Tillandsia andrieuxii* (Bromeliaceae); *Pseudoeurycea altamontana*, *P. robertsi*, *Thorius dubitus*, and *T. troglodytes* (Amphibia: Plethodontidae); *Lithobates tlaloci* (Amphibia: Ranidae); and *Romerolagus* (Mammalia: Leporidae) (Arriaga et al. 1997; Ceballos and Oliva 2005; Escalante et al. 2005; Gámez et al. 2012; Santiago-Alvarado 2019).

4.5.2.2 Orizaba-Zempoaltepec District

Orizaba-Zempoaltepec district Moore 1945, 218 (zoogeographic regionalization); Morrone 2014a, 32 (biogeographic regionalization); 2017a, 73 (biogeographic regionalization and biotic evolution); 2019, 43 (biogeographic regionalization).

Tehuacán-Cuicatlán Valley province Rzedowski 1978, 107 (phytogeographic regionalization); Rzedowski and Reyna-Trujillo 1990, map (phytogeographic regionalization).

Southern Mexico Arid district Cabrera and Willink 1973, 37 (biogeographic regionalization).

Oaxacan province (in part) Ferrusquía-Villafranca 1990, map (geomorphological regionalization).

Cañadian province Ferrusquía-Villafranca 1990, map (geomorphological regionalization).

Mixteco-Zapotecan subprovince Ferrusquía-Villafranca 1990, map (geomorphological regionalization).

Oaxaco-Tehuacanan province (in part) Ramírez-Pulido and Castro-Campillo 1990, map (zoogeographic regionalization).

Oaxacan Dry Forests ecoregion Dinerstein et al. 1995, 99 (ecological regionalization).

Oaxacan Humid Forests ecoregion Dinerstein et al. 1995, 91 (ecological regionalization).

Puebla Xeric Shrubland ecoregion Dinerstein et al. 1995, 110 (ecological regionalization).

East Veracruzan district Torres Miranda and Luna 2007, 513 (track analysis and biogeographic regionalization).

The Orizaba-Zempoaltepec district corresponds to southeastern Puebla, northern Oaxaca, and western Veracruz (Moore 1945; Rzedowski 1978). Some endemic taxa include *Oaxacania* (Asteraceae), *Solisia* (Cactaceae), *Pringleochloa* (Poaceae), *Parallocorynus bicolor* (Coleoptera: Belidae), *Pseudoeurycea leprosa* (Amphibia: Plethodontidae), *Habromys lepturus* and *Peromyscus mekisturus* (Mammalia: Cricetidae), and *Sorex macrodon* and *S. oreopolus* (Mammalia: Soricidae)

(Rzedowski 1978; Ceballos and Oliva 2005; Gámez et al. 2012; Santiago-Alvarado 2019).

4.6 Sierra Madre del Sur Province

Austral-Western province Smith 1941, 108 (zoogeographic regionalization).

Sierra Madre del Sur province Goldman and Moore 1945, 358 (zoogeographic regionalization); Moore 1945, 218 (zoogeographic regionalization); Stuart 1964, 351 (zoogeographic regionalization); Casas-Andreu and Reyna-Trujillo 1990, map (zoogeographic regionalization); Ramírez-Pulido and Castro-Campillo 1990, map (zoogeographic regionalization); Anderson and O'Brien 1996, 332 (biotic evolution); Ayala et al. 1996, 429 (zoogeographic regionalization); Arriaga et al. 1997, 65 (biogeographic regionalization); Campbell 1999, 116 (zoogeographic regionalization); Morrone et al. 1999, 510 (parsimony analysis of endemicity); Espinosa et al. 2000, 64 (parsimony analysis of endemicity and biogeographic regionalization); Morrone 2001, 39 (biogeographic regionalization); Morrone and Márquez 2001, 637 (track analysis); Morrone et al. 2002, 95 (biogeographic regionalization); Huber and Riina 2003, 263 (phytogeographic glossary); Corona and Morrone 2005, 38 (track analysis); Escalante et al. 2005, 202 (parsimony analysis of endemicity); Morrone 2005, 237 (biogeographic regionalization); Morrone and Gutiérrez 2005, 1317 (track analysis); Andrés Hernández et al. 2006, 902 (track analysis); Morrone 2006, 477 (biogeographic regionalization); Morrone and Llorente Bousquets 2006, 1021 (biogeographic regionalization); Mariño-Pérez et al. 2007, 80 (track analysis); Espinosa Organista et al. 2008, 57 (biogeographic regionalization); Morrone and Márquez 2008, 20 (track analysis); Corona et al. 2009, 1732 (track analysis); Escalante et al. 2009, 473 (endemicity analysis); León-Paniagua and Morrone 2009, 1942 (cladistic biogeography); Blancas-Calva et al. 2010, 562; Morrone 2010, 358 (biotic evolution); Escalante et al. 2011, 32 (track analysis); Coulleri and Ferrucci 2012, 105 (track analysis); Mercado-Salas et al. 2012, 459 (track analysis); Puga-Jiménez et al. 2013, 1180 (parsimony analysis of endemicity); Morrone 2014a, 33 (biogeographic regionalization); García Díaz et al. 2015, 192 (track analysis); Cuellar-Martínez and Sosa 2016, 688 (species richness and endemicity analyses); Hernández-Cerda et al. 2016, 91 (climate); Munguía-Lino et al. 2016, 146 (endemicity analysis); Santa María-Díaz 2016, 39 (geology); Santiago-Alvarado et al. 2016, 431 (biogeographic regionalization); Velasco de León et al. 2016 (paleontology); Morrone, 2017a, 74 (biogeographic regionalization and biotic evolution); 2017b (biogeographic regionalization); Morrone et al. 2017, 278 (map); Beron 2018, 630 (zoogeographic regionalization); García-Navarrete and Morrone 2018, 4 (parsimony analysis of endemicity); Rodríguez et al. 2018, 538 (species richness and endemicity); Aragón-Parada et al. 2019, 2 (diversity and richness); Estrada Sánchez et al. 2019, 133 (endemicity analysis); Morrone

2019, 46 (biogeographic regionalization); Santiago-Alvarado 2019, 39 (endemicity analysis and biogeographic regionalization).
Sierra and Mesa del Sur region West 1964, 368 (zoogeographic regionalization).
Meridional Mountain province (in part) Rzedowski 1978, 103 (phytogeographic regionalization); Rzedowski and Reyna-Trujillo 1990, map (phytogeographic regionalization); Luna Vega et al. 1999, 1301 (parsimony analysis of endemicity); Huber and Riina 2003, 260 (phytogeographic glossary).
Jaliscoan-Guerreran province Ferrusquía-Villafranca 1990, map (geomorphological regionalization).
Sierra Madre del Sur Pine-Oak Forests ecoregion Dinerstein et al. 1995, 97 (ecological regionalization).
Sierra Madre del Sur Highlands area Flores-Villela and Goyenechea 2001, 174 (cladistic biogeography).
Sierra Madre del Sur area Marshall and Liebherr 2000, 206 (cladistic biogeography); Katinas et al. 2004, 166 (parsimony analysis of endemicity).
Sierra Madre del Sur ecoregion Abell et al. 2008, 408 (ecological regionalization).
Highlands of Southern Mexico area Flores-Villela and Martínez-Salazar 2009, 820 (cladistic biogeography).

The Sierra Madre del Sur province (Fig. 4.12) comprises south central Mexico, in the states of Jalisco, Colima, Michoacán, Guerrero, Oaxaca, and parts of Puebla, at elevations above 1000 m (Morrone 2014a, 2017b; Luna-Vega et al. 2016; Santiago-Alvarado et al. 2016).

Endemic and Characteristic Taxa Morrone (2014a) provided a list of endemic and characteristic taxa. Some examples include *Montanoa grandiflora*, *M. mollissima*, and *M. tomentosa* subsp. *microcephala* (Asteraceae); *Bursera aloexylon* (Burseraceae); *Naupactus stupidus*, *Pantomorus longulus*, *Phacepholis brevipes*, and *P. globicollis* (Coleoptera: Curculionidae); *Cotinis ibarrai*, *Onthophagus bassarisus*, and *O. semiopacus* (Coleoptera: Scarabaeidae); *Abronia mixteca* and *A. oaxacae* (Squamata: Anguidae); *Cyanolyca mirabilis* (Aves: Corvidae); *Megadontomys thomasi*, *Peromyscus megalops*, and *Reithrodontomys mexicanus* (Mammalia: Cricetidae); and *Cryptotis goldmani* (Mammalia: Soricidae). Navarro-Sigüenza et al. (2016) concluded that the Sierra Madre del Sur province exhibit a high number of endemic bird species; some examples include *Cypseloides storei*, *Lophornis brachylophus*, *Eupherusa policerca*, and *E. cyanophrys* (Trochilidae), *Cyanolyca mirabilis* (Corvidae), and *Arremon kuehneri* (Passerellidae).

Vegetation Temperate forests, particularly pine-oak forests, are predominant; there are also xerophytic scrublands with cacti (Dinerstein et al. 1995). The montane forests of the Sierra Madre del Sur province represent some of the world's most diverse and complex subtropical mixed hardwood-conifer forests (Dinerstein et al. 1995). Dominant plant genera include *Achaenipodium*, *Hintonella*, *Microspermum*, *Omiltemia*, *Peyritschia*, and *Silviella* (Rzedowski 1978).

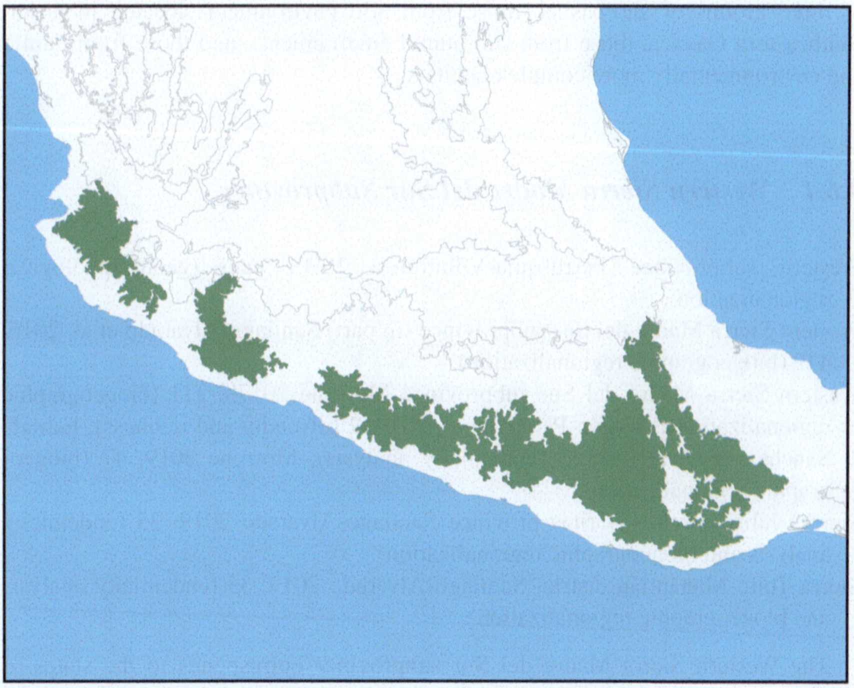

Fig. 4.12 Sierra Madre del Sur biogeographic province (Morrone et al. 2017)

Biotic Relationships According to a parsimony analysis of endemicity based on plant, insect and bird taxa (Morrone et al. 1999), the Sierra Madre del Sur province is closely related to the Pacific Lowlands, Transmexican Volcanic Belt, and Balsas Basin provinces. Another parsimony analysis of endemicity based on mammals (Escalante et al. 2005) found that it is related to the Chiapas Highlands province and other areas assigned to the Neotropical region. A track analysis based on species of Coleoptera (Morrone and Márquez 2001) showed that this province is related to the Sierra Madre Occidental, Balsas Basin, Sierra Madre Oriental, and Transmexican Volcanic Belt provinces. Cladistic biogeographic analyses (Marshall and Liebherr 2000; Miguez-Gutiérrez et al. 2013) postulated a close relationship with the Transmexican Volcanic Belt province.

Regionalization Some nested units that have been identified within this province (Smith 1941; Ferrusquía-Villafranca 1990; Arriaga et al. 1997; Escalante et al. 1998; Santiago-Alvarado et al. 2016) were treated by Morrone (2017b, 2019) as three subprovinces and five districts.

Blancas-Calva et al. (2010) analyzed the relationships of 26 sub-basins of the Sierra Madre del Sur province, based on distributional bird data, applying a parsimony analysis of endemicity (PAE). The cladogram obtained suggests the existence of three groups of sub-basins: those from arid environments located in north-southeastern Oaxaca, those from sub-humid environments, and those from humid and environmentally more complex habitats.

4.6.1 Western Sierra Madre del Sur Subprovince

Western subprovince Ferrusquía-Villafranca 1990, map (geomorphological regionalization).
Western Sierra Madre del Sur subprovince (in part) Santiago-Alvarado et al. 2016, 439 (biogeographic regionalization).
Western Sierra Madre del Sur subprovince Morrone 2017b, 711 (biogeographic regionalization); Aragón-Parada et al. 2019, 2 (diversity and richness); Estrada Sánchez et al. 2019, 135 (endemicity analysis); Morrone 2019, 47 (biogeographic regionalization).
Sierras Jaliscienses-Nayaritas province Santiago-Alvarado 2019, 35 (endemicity analysis and biogeographic regionalization).
Sierra Tuito-Manantlán district Santiago-Alvarado 2019, 35 (endemicity analysis and biogeographic regionalization).

The Western Sierra Madre del Sur subprovince corresponds to the states of Jalisco, Colima, and Michoacán. Supported by *Pinus jaliscana* (Pinaceae); *Psacalium multilobum*, *P. pentaflorum*, and *P. poculiferum* (Asteraceae); *Quercus aristata* (Fagaceae); and *Lithobates pustulosus* (Amphibia: Ranidae) (Santiago-Alvarado 2019). It comprises two districts.

4.6.1.1 Jaliscian District

Jaliscian district Moore 1945, 218 (zoogeographic regionalization); Morrone 2014a. 32 (biogeographic regionalization); 2017a, 72 (biogeographic regionalization and biotic evolution); 2017b, 711 (biogeographic regionalization); Aragón-Parada et al. 2019, 2 (diversity and richness); Morrone 2019, 47 (biogeographic regionalization).
West district Torres Miranda and Luna 2007, 512 (track analysis and biogeographic regionalization).
Jaliscian-Tuito district Santiago-Alvarado et al. 2016, 439 (biogeographic regionalization).
Jaliscan district García-Navarrete and Morrone 2018, 11 (parsimony analysis of endemicity).
Jalisciense district Estrada Sánchez et al. 2019, 135 (endemicity analysis).

The Jaliscian district is situated in the northern part of the subprovince, in the state of Jalisco. Supported by *Quercus cualensis* and *Q. tuitensis* (Fagaceae), and

Rhabdias manantlanensis (Nematoda: Rhabdiasidae) (Morrone 2017b; Santiago-Alvarado 2019).

4.6.1.2 Jaliscian-Manantlán District

Jaliscian-Manantlán district Santiago-Alvarado et al. 2016, 440 (biogeographic regionalization); Morrone 2017b, 711 (biogeographic regionalization); Aragón-Parada et al. 2019, 2 (diversity and richness); Morrone 2019, 47 (biogeographic regionalization).

Jaliscan-Manantlán district García-Navarrete and Morrone 2018, 11 (parsimony analysis of endemicity).

Jalisciense-Manantlán district Estrada Sánchez et al. 2019, 135 (endemicity analysis).

The Jaliscian-Manantlán district is situated in the southern part of the subprovince, in the states of Jalisco, Michoacán, and northern Colima. Supported by *Beilschmiedia manantlanensis* (Lauraceae), *Populus guzmanantlanensis* (Salicaceae), and *Canthon occidentalis* (Coleoptera: Scarabaeidae) (Morrone 2017b).

4.6.2 *Central Sierra Madre del Sur Subprovince*

Western Sierra Madre del Sur subprovince (in part) Santiago-Alvarado et al. 2016, 439 (biogeographic regionalization).

Central Sierra Madre del Sur subprovince Morrone 2017b, 711 (biogeographic regionalization); Morrone 2019, 47 (biogeographic regionalization).

The Central Sierra Madre del Sur subprovince corresponds to the central part of the province, in the state of Michoacán. It comprises a single district.

4.6.2.1 Michoacán District

Michoacán district Santiago-Alvarado et al. 2016, 440 (biogeographic regionalization); Morrone 2017b, 711 (biogeographic regionalization); 2019, 47 (biogeographic regionalization).

Coalcomán province Santiago-Alvarado 2019, 39 (endemicity analysis and biogeographic regionalization).

The Michoacán district has the same circumscription than the province. Supported by *Pinus rzedowskii* (Pinaceae), *Bursera confusa* (Burseraceae), and *Peromyscus sagax* and *P. winkelmanni* (Mammlia: Cricetidae) (Morrone 2017b; Santiago-Alvarado 2019).

4.6.3 Eastern Sierra Madre del Sur Subprovince

Eastern Sierra Madre del Sur subprovince Santiago-Alvarado et al. 2016, 440 (biogeographic regionalization); Morrone 2017b, 711 (biogeographic regionalization); Aragón-Parada et al. 2019, 2 (diversity and richness); Morrone 2019, 47 (biogeographic regionalization).

The Eastern Sierra Madre del Sur subprovince corresponds to the eastern portion of the province, in the states of Guerrero, Oaxaca, and part of Puebla. It comprises two districts.

4.6.3.1 Guerreran District

Guerreran province Smith 1941, 108 (zoogeographic regionalization); Ramírez-Pulido and Castro-Campillo 1990, map (zoogeographic regionalization); Brown et al. 1998, 30 (vegetation).
Sierra Madre del Sur-Guerrero province Escalante et al. 1998, 284 (cluster analysis and biogeographic regionalization).
Guerreran district Morrone 2014a, 33 (biogeographic regionalization); 2017a, 74 (biogeographic regionalization and biotic evolution); 2017b, 712 (biogeographic regionalization); Aragón-Parada et al. 2019, 2 (diversity and richness); Morrone 2019, 48 (biogeographic regionalization).
Sierras Guerrerenses-Sierra Mixteca district Santiago-Alvarado 2019, 40 (endemicity analysis and biogeographic regionalization).

The Guerreran district corresponds to the westernmost part of the Sierra Madre del Sur province, in the state of Guerrero. Supported by *Tigridia hintonii* (Iridaceae), *Quercus rubramenta* (Fagaceae), *Elliptoleus whiteheadi* (Coleoptera: Carabidae), *Cotinis ibarrai* and *Onthophagus bassariscus* (Coleoptera: Scarabaeidae), *Lithobates sierramadrensis* (Amphibia: Ranidae), *Mesaspis gadovii gadovii* (Squamata: Anguidae), *Lophornis brachylophus* (Aves: Trochilidae), *Handleyomys guerrerensis* and *Reithrodontomys bakeri* (Coleoptera: Cricetidae), and *Sylvilagus insonus* (Coleoptera: Leporidae) (Morrone 2017b; Santiago-Alvarado 2019).

4.6.3.2 Oaxacan Highlands District

Oaxacan Highlands province Smith 1941, 107 (zoogeographic regionalization).
Valles-Centralian subprovince Ferrusquía-Villafranca 1990, map (geomorphological regionalization).
Isthmic subprovince Ferrusquía-Villafranca 1990, map (geomorphological regionalization).
Sierra de Juárez Pine-Oak Forests ecoregion Dinerstein et al. 1995, 102 (ecological regionalization).

Oaxaca province Arriaga et al. 1997, 65 (biogeographic regionalization); Espinosa Organista et al. 2008, 57 (biogeographic regionalization); García Díaz et al. 2015, 194 (track analysis).
Nudo de Zempoaltépetl province Escalante et al. 1998, 284 (cluster analysis and biogeographic regionalization).
Sierra Madre del Sur-Oaxaca province Escalante et al. 1998, 284 (cluster analysis and biogeographic regionalization).
Sierra de Miahuatlán province Escalante et al. 1998, 284 (cluster analysis and biogeographic regionalization).
Central Valleys district Morrone 2014a, 33 (biogeographic regionalization); 2017a, 76 (biogeographic regionalization and biotic evolution).
Isthmian district Morrone 2014a, 33 (biogeographic regionalization); 2017a, 76 (biogeographic regionalization and biotic evolution).
Nudo de Zempoaltépetl district Morrone 2014a, 33 (biogeographic regionalization); 2017a, 76 (biogeographic regionalization and biotic evolution).
Sierra de Miahuatlán district Morrone 2014a, 33 (biogeographic regionalization); 2017a, 76 (biogeographic regionalization and biotic evolution); Santiago-Alvarado 2019, 40 (endemicity analysis and biogeographic regionalization).
Oaxacan Highlands district Morrone 2014a, 33 (biogeographic regionalization); 2017a, 76 (biogeographic regionalization and biotic evolution); 2017b, 712 (biogeographic regionalization), Aragón-Parada et al. 2019, 2 (diversity and richness); Estrada Sánchez et al. 2019, 136 (endemicity analysis); Morrone 2019, 48 (biogeographic regionalization).
Oaxacan district Santiago-Alvarado et al. 2016, 442 (biogeographic regionalization).
Sierras Circum-Valles Centrales de Oaxaca subprovince Santiago-Alvarado 2019, 39 (endemicity analysis and biogeographic regionalization).
Sierra de Juárez district Santiago-Alvarado 2019, 40 (endemicity analysis and biogeographic regionalization).
Sierra Triqui-Mixteca district Santiago-Alvarado 2019, 40 (endemicity analysis and biogeographic regionalization).
Sierra de Juárez-Sierra Mixteca nodal district Santiago-Alvarado 2019, 42 (endemicity analysis and biogeographic regionalization).

The Oaxacan Highlands district is situated in the state of Oaxaca, mainly corresponding to the Sierra Mixteca and Sierra Norte. Supported by *Ceratozamia mixeorum*, *C. whitelockiana*, and *Dioon holmgrenii* (Zamiaceae); *Bursera altijuga*, *B. arida*, *B. biflora*, and *B. heliae* (Burseraceae); *Weberocereus alliodorus* (Cactaceae); *Elliptoleus crepericornis* and *E. zapotecorum* (Coleoptera: Carabidae); *Phacepholis globicollis* (Coleoptera: Curculionidae); *Pseudoeurycea smithi* and *P. unguidentis* (Amphibia: Plethodontidae); *Habromys chinanteco* and *Microtus oaxacensis* (Mammalia: Cricetidae); and *Cryptotis peregrina* and *C. phillipsii* (Mammalia: Soricidae) (Contreras-Medina 2016; Morrone 2017b; Santiago-Alvarado 2019). This district includes practically all the vegetation types (Espinosa Organista et al. 2008).

4.7 Chiapas Highlands Province

Chiapas Highlands province Smith 1941, 109 (zoogeographic regionalization); Goldman and Moore 1945, 359 (zoogeographic regionalization); Arriaga et al. 1997, 66 (biogeographic regionalization); Espinosa et al. 2000, 64 (parsimony analysis of endemicity and biogeographic regionalization); Huber and Riina 2003, 103 (phytogeographic glossary); Espinosa Organista et al. 2008, 58 (track analysis and biogeographic regionalization); Morrone 2014a, 34 (biogeographic regionalization); García Díaz et al. 2015, 194 (track analysis); Cuellar-Martínez and Sosa 2016, 688 (species richness and endemicity analyses); Morrone 2017a, 77 (biogeographic regionalization and biotic evolution); Morrone et al. 2017, 278 (map); García-Navarrete and Morrone 2018, 4 (parsimony analysis of endemicity); Rodríguez et al. 2018, 538 (species richness and endemicity); Morrone 2019, 48 (biogeographic regionalization).

Chiapas province Barrera 1962, 101 (zoogeographic regionalization); Ferrusquía-Villafranca 1990, map (geomorphological regionalization); Ramírez-Pulido and Castro-Campillo 1990, map (zoogeographic regionalization); Morrone et al. 1999, 510 (parsimony analysis of endemicity); Morrone 2001, 45 (biogeographic regionalization); Morrone and Márquez 2001, 637 (track analysis); Morrone et al. 2002, 99 (biogeographic regionalization); Escalante et al. 2003, 570 (biogeographic regionalization); Escalante et al. 2005, 202 (parsimony analysis of endemicity); Morrone and Gutiérrez 2005, 1316 (track analysis); Morrone 2006, 478 (biogeographic regionalization); Quijano-Abril et al. 2006, 1269 (track analysis); Mariño-Pérez et al. 2007, 80 (track analysis); Morrone and Márquez 2008, 20 (track analysis); Corona et al. 2009, 1732 (track analysis); Escalante et al. 2009, 473 (endemicity analysis); Morrone 2010, 358 (biotic evolution); Escalante et al. 2011, 32 (track analysis); Mercado-Salas et al. 2012, 459 (track analysis).

Chiapas-Guatemalan Highlands province Ryan 1963, 23 (zoogeographic regionalization); Stuart 1964, 357 (zoogeographic regionalization); Miller 1966, 783 (zoogeographic regionalization); Campbell 1999, 116 (zoogeographic regionalization).

Hondurean Upland province Savage 1966, 736 (biotic evolution).

Central American Montane Forest center Müller 1973, 14 (zoogeographic regionalization).

Transisthmic Mountain province Rzedowski 1978, 103 (phytogeographic regionalization); Rzedowski and Reyna-Trujillo 1990, map (phytogeographic regionalization); Ayala et al. 1996, 429 (zoogeographic regionalization); Luna Vega et al. 1999, 1301 (parsimony analysis of endemicity); Huber and Riina 2003, 260 (phytogeographic glossary).

Sierra Madre de Chiapas province Casas-Andreu and Reyna-Trujillo 1990, map (zoogeographic regionalization); Anderson and O'Brien 1996, 332 (biotic evolution).

Central American Montane Forests ecoregion Dinerstein et al. 1995, 87 (ecological regionalization).
Central American Pine-oak Forests ecoregion Dinerstein et al. 1995, 97 (ecological regionalization).
Sierra Madre Moist Forests ecoregion Dinerstein et al. 1995, 87 (ecological regionalization).
Sierra Norte de Chiapas province Escalante et al. 1998, 285 (cluster analysis and biogeographic regionalization)
Chiapan/Guatemalan Highlands area Marshall and Liebherr 2000, 206 (cladistic biogeography).
Highlands of Chiapas and Guatemala area Flores-Villela and Goyenechea 2001, 174 (cladistic biogeography); Flores-Villela and Martínez-Salazar 2009, 820 (cladistic biogeography).
Highlands of Northern Central America region Huber and Riina 2003, 156 (phytogeographic glossary).
Chiapas-Fonseca ecoregion Abell et al. 2008, 408 (ecological regionalization).
Chortís block province (in part) Townsend 2014, 206 (paleogeography).
Central American Cordillera province Santiago-Alvarado 2019, 41 (endemicity analysis and biogeographic regionalization).
Chiapas Highlands-Sierra Los Cuchumatanes subprovince Santiago-Alvarado 2019, 41 (endemicity analysis and biogeographic regionalization).

The Chiapas Highlands biogeographic province (Fig. 4.13) comprises southern Mexico, Guatemala, Honduras, El Salvador, and Nicaragua and basically corresponds to the Central American Nucleus, from 500 to 2000 m altitude (Morrone 2006, 2014a). Marshall and Liebherr (2000) combined the Chiapas Highlands and Yucatán Peninsula provinces in a single province, based on their shared taxa.

Endemic and Characteristic Taxa Espinosa Organista et al. (2008) and Morrone (2014a) provided lists of endemic and characteristic taxa. Some examples include *Elaphoglossum latum* (Dryopteridaceae); *Polypodium chiapense* (Polypodaceae); *Juniperus comitana* (Cupressaceae); *Cecropia sylvicola* (Cecropiaceae); *Tetranema evolutum* (Scrophulariaceae); *Acropsopilio chomulae* (Opiliones: Caddidae); *Toonglasa indomita* (Araneae: Lygaeidae); *Baronia brevicornis rufodiscalis* and *Priamides erostratus erostratus* (Lepidoptera: Papilionidae); *Bledius strenuus*, *Gansia andersoni*, and *Styngetus championi* (Coleoptera: Staphylinidae); *Hyla euphorbiacea biseriata* (Amphibia: Hylidae); *Sceloporus malachiticus* (Squamata: Phrynosomatidae); *Bothriechis aurifer* and *B. bicolor* (Squamata: Viperidae); and *Microtus guatemalensis*, *Oryzomys rhabdops*, *O. saturatior*, *Ototylomys phyllotis connectens*, *Peromyscus guatemalensis*, and *Tylomys bullaris* (Mammalia: Cricetidae).

Vegetation Different types of temperate forests (Fig. 4.14), with species of *Pinus* and *Quercus* being dominant, savannas, and scrublands (Rzedowski 1978; Dinerstein

Fig. 4.13 Chiapas Highlands biogeographic province (map by Gerardo Rodríguez-Tapia)

et al. 1995; Meave et al. 2016). The pine-oak forest is dominated by *Pinus ayacahuite*, *P. montezumae*, *P. pseudostrobus*, *P. oocarpa*, *Quercus benthamii*, *Q. crassifolia*, *Q. crispipilis*, and *Q. rugosa* (Martínez-Icó et al. 2015).

Biotic Relationships A parsimony analysis of endemicity based on plant, insect, and bird taxa (Morrone et al. 1999) postulated that this province is closely related to the Veracruzan province. Another parsimony analysis of endemicity based on mammals (Escalante et al. 2005) found that it is related to other areas from the Neotropical region. A track analysis based on species of Coleoptera (Morrone and Márquez 2001) showed that this province is related to the Puntarenas-Chiriquí, Veracruzan, and Pacific Lowlands provinces.

Regionalization Some nested units that have been identified within the Chiapas Highlands biogeographic province (Ryan 1963; Stuart 1964; Müller 1973; Rzedowski 1978; Ferrusquía-Villafranca 1990; Rzedowski and Reyna-Trujillo 1990) were treated by Morrone (2014a, 2019) as six districts; their precise delimitation is not without doubt.

4.7.1 *Sierra Madrean District*

Sierra-Madrean subprovince Ferrusquía-Villafranca 1990, map (geomorphological regionalization).
Sierra Madre de Chiapas province Escalante et al. 1998, 284 (cluster analysis and biogeographic regionalization).
Sierra Madrean district Morrone 2014a, 35 (biogeographic regionalization); 2017a, 79 (biogeographic regionalization and biotic evolution); 2019, 49 (biogeographic regionalization).
Sierra Madre de Chiapas district Santiago-Alvarado 2019, 42 (endemicity analysis and biogeographic regionalization).

The Sierra Madrean district corresponds to the western portion of the Chiapas Highlands province, in eastern Oaxaca (Ferrusquía-Villafranca 1990). Some endemic taxa include *Pseudoeurycea brunnata* and *P. goebeli* (Amphibia: Plethodontidae), *Abronia matudai* (Squamata: Anguidae), *Habromys lophurus* and *Tylomys tumbalensis* (Mammalia: Cricetidae), *Heteromys nelsoni* (Mammalia: Heteromyidae), and *Cryptotis griseoventris* (Mammalia: Soricidae) (Escalante et al. 2005; Santiago-Alvarado 2019).

4.7.2 *Comitanian District*

Comitanian subprovince Ferrusquía-Villafranca 1990, map (geomorphological regionalization).
Comitanian district Morrone 2014a, 35 (biogeographic regionalization); 2017a, 79 (biogeographic regionalization and biotic evolution); 2019, 49 (biogeographic regionalization).

The Comitanian district corresponds to the southern portion of the Chiapas Highlands province, in eastern Oaxaca and Chiapas (Ferrusquía-Villafranca 1990).

4.7.3 *Lacandonian District*

Lacandonian subprovince Ferrusquía-Villafranca 1990, map (geomorphological regionalization).
Lacandonian district Morrone 2014a, 35 (biogeographic regionalization); 2017a, 79 (biogeographic regionalization and biotic evolution); 2019, 49 (biogeographic regionalization).

The Lacandonian district corresponds to the northern portion of the Chiapas Highlands province, in Chiapas (Ferrusquía-Villafranca 1990). An endemic taxon is *Cryptotis griseoventris* (Mammalia: Soricidae).

Fig. 4.14 Vegetation of the Chiapas Highlands province, Los Chimalapas, Oaxaca. (**a**) Cloud forest; (**b**) pine-oak forest (pictures by Víctor Moctezuma)

4.7.4 Soconusco District

Soconusco province Rzedowski 1978, 109 (phytogeographic regionalization); Rzedowski and Reyna-Trujillo 1990, map (phytogeographic regionalization); Arriaga et al. 1997, 64 (biogeographic regionalization); Espinosa Organista et al. 2008, 58 (biogeographic regionalization).

Soconusco district Morrone 2014a, 35 (biogeographic regionalization); 2017a. 79 (biogeographic regionalization and biotic evolution); 2019, 49 (biogeographic regionalization).

The Soconusco district is a narrow strip in the lower foothills of the Sierra Madre de Chiapas, Mexico, and part of Guatemala (Rzedowski 1978). Endemic taxa include *Asplenium solmsii* (Aspleniaceae), *Zamia soconuscensis* (Zamiaceae), *Quercus durantifolia* (Fagaceae), and *Abronia smithi* (Squamata: Anguidae) (Espinosa Organista et al. 2008).

4.7.5 Guatemalan Highland District

Guatemalan Highland district Morrone 2014a, 35 (biogeographic regionalization); 2017a, 79 (biogeographic regionalization and biotic evolution).

Sierra Los Cuchumatanes nodal district Santiago-Alvarado 2019, 43 (endemicity analysis and biogeographic regionalization).

The Guatemalan Highland district corresponds to the highlands of Guatemala, in the Sierra Los Cuchumatanes (Santiago-Alvarado 2019). Endemic taxa include *Bolitoglossa hartwegi, B. lincolni, and B. rostrata* (Amphibia: Plethodontidae) and *Peromyscus zarhynchus* and *P. guatemalensis* (Mammalia: Cricetidae) (Santiago-Alvarado 2019).

4.7.6 Nicaraguan Montane District

Nicaraguan Montane province Ryan 1963, 28 (zoogeographic regionalization).

Honduran-Nicaraguan Highlands province Stuart 1964, 357 (zoogeographic regionalization).

Nicaraguan Montane district Morrone 2014a, 35 (biogeographic regionalization); Morrone 2017a, 79 (biogeographic regionalization and biotic evolution); García-Navarrete and Morrone 2018, 11 (parsimony analysis of endemicity)

The Nicaraguan Montane district corresponds to the highlands of Nicaragua (Ryan 1963).

References

Abell R, Thieme ML, Revenga C, Bryer M, Kottelat M, Bogutskaya N, Coad B, Mandrak N, Contreras Balderas S, Bussing W, Stiassny MLL, Skelton P, Allen GR, Unmack P, Naseka A, Ng R, Sinforf N, Robertson J, Armijo E, Higgins JV, Heibel TJ, Wikramanayake E, Olson D, López HL, Reis RE, Lundberg JG, Sabaj Pérez MH, Petry P (2008) Freshwater ecoregions of the world: a new map of biogeographic units for freshwater biodiversity conservation. Bioscience 58:493–414

Alcántara O, Paniagua M (2007) Patrones de distribución y conservación de plantas endémicas. In: Luna I, Morrone JJ, Espinosa D (eds) Biodiversidad de la Faja Volcánica Transmexicana, Las Prensas de Ciencias. UNAM, Mexico City, pp 421–447

Anderson RS, O'Brien CW (1996) Curculionidae (Coleoptera). In: Llorente Bousquets J, García Aldrete AN, Gónzalez Soriano E (eds) Biodiversidad, taxonomía y biogeografía de artrópodos de México: Hacia una síntesis de su conocimiento. Universidad Nacional Autónoma de México, Mexico City, pp 329–351

Andrés Hernández AR, Morrone JJ, Terrazas T, López Mata L (2006) Análisis de trazos de las especies mexicanas de *Rhus* subgénero *Lobadium* (Angiospermae: Anacardiaceae). Interciencia 31:900–904

Aragón-Parada J, Carrillo-Reyes P, Rodríguez A, Munguía-Lino G (2019) Diversidad y distribución geográfica del género *Sedum* (Crassulaceae) en la Sierra Madre del Sur, México. Rev Mex Biodivers 90:e90291

Arbeláez-Cortés E, Nyari AS, Navarro-Sigüenza AG (2010) The differential effect of lowlands on the phylogeographic pattern of a Mesoamerican montane species (*Lepidocolaptes affinis*, Aves: Furnariidae). Mol Phylogenet Evol 57:658–668

Arellano E, González-Cozátl FX, Rogers DS (2005) Molecular systematics of Middle American harvest mice *Reithrodontomys* (Muridae), estimated from mitochondrial cytochrome b gene sequences. Mol Phylogenet Evol 37:529–540

Arriaga L, Aguilar C, Espinosa D, Jiménez R (eds) (1997) Regionalización ecológica y biogeográfica de México. Taller de la Comisión Nacional para el Conocimiento y Uso de la Biodiversidad (Conabio), November 1997, Mexico City

Arriaga-Jiménez A, Rös M, Halffter G (2018) High variability of dung beetle diversity patterns at four mountains of the trans-Mexican Volcanic Belt. PeerJ 6:e4468

Ayala R, Griswold TL, Yanega D (1996) Apoidea (Hymenoptera). In: Llorente Bousquets J, García Aldrete AN, Gónzalez Soriano E (eds) Biodiversidad, taxonomía y biogeografía de artrópodos de México: Hacia una síntesis de su conocimiento. Universidad Nacional Autónoma de México, Mexico City, pp 423–464

Barrera A (1962) La península de Yucatán como provincia biótica. Rev Soc Mex Hist Nat 23:71–105

Barrera-Moreno O, Escalante T, Rodríguez G (2011) Panbiogeografía y modelos digitales de elevación: Un caso de estudio con roedores en la Faja Volcánica Transmexicana. Rev Geogr Norte Grande 48:11–25

Bartholomew JG, Clark WE, Grimshaw PH (1911) Atlas of zoogeography: a series of maps illustrating the distribution of over seven hundred families, genera, and species of existing animals. Edinburgh Geographical Institute, Edinburgh

Beron P (2018) Zoogeography of Arachnida. Springer, Cham

Blancas-Calva E, Navarro-Sigüenza AG, Morrone JJ (2010) Patrones biogeográficos de la avifauna de la Sierra Madre del Sur. Rev Mex Biodivers 81:561–568

Brown DE, Reichenbacher F, Franson SE (1998) A classification of north American biotic communities. The University of Utah Press, Salt Lake City

Cabrera AL, Willink A (1973) Biogeografía de América Latina. Monografía 13, Serie de Biología, OEA, Washington, DC

Campbell JA (1999) Distribution patterns of amphibians in Middle America. In: Duellman WE (ed) Patterns of distribution of amphibians: a global perspective. The Johns Hopkins University Press, Baltimore and London, pp 111–210

Canseco-Márquez L, Mendoza-Quijano F, Gutiérrez-Mayén MG (2004) Análisis de la distribución de la herpetofauna. In: Luna I, Morrone JJ, Espinosa D (eds) Biodiversidad de la Sierra Madre Oriental, Las Prensas de Ciencias. UNAM, Mexico City, pp 417–437

Carleton MD, Sánchez O, Urbano-Vidales G (2002) A new species of *Habromys* (Muroidea: Neotominae) from Mexico, with a generic review of species definitions and remarks on diversity patterns among Mesoamerican small mammals restricted to humid montane forests. Proc Biol Soc Washington 115:488–533

Casas-Andreu G, Reyna-Trujillo T (1990) Herpetofauna (anfibios y reptiles). Mapa IV. 8. 6. In: Atlas Nacional de México, vol III, Instituto de Geografía, UNAM, Mexico City, map

Castillo C, Reyes-Castillo P (1984) Biosistemática del género *Petrejoides* Kuwert (Coleoptera, Lamellicornia, Passalidae). Acta Zool Mex (n. s.) 4:1–84

Ceballos G, Oliva G (2005) Los mamíferos silvestres de México. Conabio and Fondo de Cultura Económica, Mexico City

Challenger A (1988) Utilización y conservación de los ecosistemas terrestres de México: Pasado, presente y futuro. Conabio, Instituto de Biología-UNAM and Agrupación Sierra Madre S.C., Mexico City

Clarke CB (1892) On biologic regions and tabulation areas. Phil Trans R Soc Lond 183:371–387

Conroy CJ, Hortelano Y, Cervantes FA, Cook JA (2001) The phylogenetic position of southern relictual species of *Microtus* (Muridae: Rodentia) in North America. Mamm Biol 56:332–344

Contreras-Medina R (2016) Las gimnospermas de la Sierra Madre del Sur. In: Luna-Vega I, Espinosa D, Contreras-Medina R (eds) Biodiversidad de la Sierra Madre del Sur: Una síntesis preliminar. UNAM, Mexico City, pp 157–165

Contreras-Medina R, Luna Vega I, Morrone JJ (2007a) Gymnosperms and cladistic biogeography of the Mexican transition zone. Taxon 56:905–915

Contreras-Medina R, Luna-Vega I, Morrone JJ (2007b) Application of parsimony analysis of endemicity to Mexican gymnosperm distributions: grid-cells, biogeographical provinces and track analysis. Biol J Linn Soc 92:405–417

Corona AM, Morrone JJ (2005) Track analysis of the species of *Lampetis (Spinthoptera)* Casey, 1909 (Coleoptera: Buprestidae) in North America, Central America, and the West Indies. Caribb J Sci 41:37–41

Corona AM, Toledo VH, Morrone JJ (2007) Does the trans-Mexican Volcanic Belt represent a natural biogeographic unit?: An analysis of the distributional patterns of Coleoptera. J Biogeogr 34:1008–1015

Corona AM, Toledo VH, Morrone JJ (2009) Track analysis of the Mexican species of Buprestidae (Coleoptera): testing the complex nature of the Mexican transition zone. J Biogeogr 36:1730–1738

Corral-Rosas V, Morrone JJ (2017) Analyzing the assembly of cenocrons in the Mexican transition zone through a time-sliced cladistic biogeographic analysis. Aust Syst Bot 29:489–501

Coulleri JP, Ferrucci MS (2012) Biogeografía histórica de *Cardiospermum* y *Urvillea* (Sapindaceae) en América: Paralelismos geográficos e históricos con los bosques secos estacionales neotropicales. Bol Soc Argent Bot 47:103–117

Cuellar-Martínez M, Sosa V (2016) Diversity patterns of monocotiledonous geophytes in Mexico. Bot Sci 94:687–699

Darlington PJ (1957) Zoogeography: the geographical distribution of animals. Wiley, New York

de la Rosa-Manzano E, Mendieta-Leiva G, Guerra-Pérez A, Aguilar-Dorantes KM, Arellano-Méndez LU, Torres-Castillo JA (2019) Vascular epiphytic diversity in a neotropical transition zone is driven by environmental and structural heterogeneity. Trop Conserv Sci 12:1–16

de Mello-Leitão C (1937) Zoo-geografia do Brasil. Biblioteca Pedagógica Brasileira, Brasiliana, São Paulo

Dinerstein E, Olson DM, Graham DJ, Webster AL, Primm SA, Bookbinder MP, Ledec G (1995) A conservation assessment of the terrestrial ecoregions of Latin America and the Caribbean. The World Bank, Washington, DC

Domínguez-Domínguez O, Vila M, Pérez-Rodríguez R, Remón N, Doadrio I (2011) Complex evolutionary history of the Mexican stoneroller *Campostoma ornatum* Girard, 1856 (Actinopterygii: Cyprinidae). BMC Evol Biol 11:1–20

Edwards CW, Bradley RD (2002a) Molecular systematics and historical phylobiogeography of the *Neotoma mexicana* species group. J Mammal 83:20–30

Edwards CW, Bradley RD (2002b) Molecular systematics of the genus *Neotoma*. Mol Phylogenet Evol 25:489–500

Engler A (1882) Versuch einer Entwicklungsgeschichte der Pflanzenwelt, insbesondere der Florengebiete seit der Tertiärperiode. Vol. 2. Die extratropischen Gebiete der Südlichen Hemisphäre und die Tropischen Gebiete. Verlag von W. Engelmann, Leipzig

Escalante P, Navarro AG, Peterson AT (1998) Un análisis geográfico, ecológico e histórico de la diversidad de aves terrestres de México. In: Ramamoorthy TP, Bye R, Lot A, Fa A (eds) Diversidad biológica de México: Orígenes y distribución. Instituto de Biología, UNAM, Mexico City, pp 279–304

Escalante T, Espinosa D, Morrone JJ (2003) Using parsimony analysis of endemicity to analyze the distribution of Mexican land mammals. Southwest Nat 48:563–578

Escalante T, Rodríguez G, Morrone JJ (2004) The diversification of Nearctic mammals in the Mexican transition zone. Biol J Linn Soc 83:327–339

Escalante T, Rodríguez G, Morrone JJ (2005) Las provincias biogeográficas del Componente Mexicano de Montaña desde la perspectiva de los mamíferos continentales. Rev Mex Biodivers 76:199–205

Escalante T, Rodríguez G, Cao N, Ebach MC, Morrone JJ (2007a) Cladistic biogeographic analysis suggests an early Caribbean diversification in Mexico. Naturwissenschaften 94:561–565

Escalante T, Rodríguez G, Gámez N, León-Paniagua L, Barrera O, Sánchez-Cordero V (2007b) Biogeografía y conservación de los mamíferos. In: Luna I, Morrone JJ, Espinosa D (eds) Biodiversidad de la Faja Volcánica Transmexicana. Las Prensas de Ciencias, UNAM, Mexico City, pp 485–502

Escalante T, Szumik C, Morrone JJ (2009) Areas of endemism of Mexican mammals: reanalysis applying the optimality criterion. Biol J Linn Soc 98:468–478

Escalante T, Martínez-Salazar EA, Falcón-Ordaz J, Linaje M, Guerrero R (2011) Análisis panbiogeográfico de *Vexillata* (Nematoda: Ornithostrongylidae) y sus huéspedes (Mammalia: Rodentia). Acta Zool Mex (n. s.) 27:25–46

Espinosa D, Ocegueda S (2007) Introducción. In: Luna I, Morrone JJ, Espinosa D (eds) Biodiversidad de la Faja Volcánica Transmexicana. Las Prensas de Ciencias, UNAM, Mexico City, pp 5–6

Espinosa D, Morrone JJ, Aguilar C, Llorente J (2000) Regionalización biogeográfica de México: Provincias bióticas. In: Llorente J, González E, Papavero N (eds) Biodiversidad, taxonomía y biogeografía de artrópodos de México: Hacia una síntesis de su conocimiento, vol II. UNAM, Mexico City, pp 61–94

Espinosa D, Aguilar C, Ocegueda S (2004) Identidad biogeográfica de la Sierra Madre Oriental y posibles subdivisiones bióticas. In: Luna I, Morrone JJ, Espinosa D (eds) Biodiversidad de la Sierra Madre Oriental. Las Prensas de Ciencias, UNAM, Mexico City, pp 487–500

Espinosa Organista D, Ocegueda Cruz S, Aguilar Zúñiga C, Flores Villela Ó, Llorente-Bousquets J (2008) El conocimiento biogeográfico de las especies y su regionalización natural. In: Sarukhán J (ed) Capital natural de México, Conocimiento actual de la biodiversidad, vol I. Conabio, Mexico City, pp 33–65

Estrada Sánchez I, García-Cruz J, Espejo-Serna A, López-Ortega G (2019) Identification of areas of endemism in the Mexican cloud forests based on the distribution of endemic epiphytic bromeliads and orchids. Phytotaxa 397:129–145

Ferrari L, Orozco-Esquivel T, Manea V, Manea M (2012) The dynamic history of the trans-Mexican volcanic belt and the Mexico subduction zone. Tectonophysics 522–523:122–149

Ferro I, Navarro-Sigüenza AG, Morrone JJ (2017) Biogeographical transitions in the Sierra Madre oriental, Mexico, shown by chorological and evolutionary biogeographical affinities of Passerine birds (Aves: Passeriformes). J Biogeogr 44:2145–2160

Ferrusquía-Villafranca I (1990) Regionalización biogeográfica. Mapa IV. 8. 10. In: Atlas Nacional de México, vol III. Instituto de Geografía, UNAM, Mexico City, map

Ferrusquía-Villafranca I (1998) Geología de México: Una sinopsis. In: Ramamoorty TP, Bye R, Lot A, Fa J (eds) Diversidad biológica de México: Orígenes y distribución. Instituto de Biología, UNAM, Mexico City, pp 3–108

Ferrusquía-Villafranca I (2007) Ensayo sobre la caracterización y significación biológica. In: Luna I, Morrone JJ, Espinosa D (eds) Biodiversidad de la Faja Volcánica Transmexicana. Las Prensas de Ciencias, UNAM, Mexico City, pp 7–23

Ficetola GF, Mazel F, Thuiller W (2017) Global determinants of zoogeographic boundaries. Nat Ecol Evol 1:1–7

Flores-Villela O, Goyenechea I (2001) A comparison of hypotheses of historical area relationships for Mexico and Central America, or in search for the lost pattern. In: Johnson J, Webb RG, Flores-Villela O (eds) Mesoamerican herpetology: systematics, zoogeography, and conservation, Centennial Museum, Special Publication 1. University of Texas, El Paso, pp 171–181

Flores-Villela O, Martínez-Salazar EA (2009) Historical explanation of the origin of the herpetofauna of México. Rev Mex Biodivers 80:817–833

Gámez N, Escalante T, Rodríguez G, Linaje M, Morrone JJ (2012) Caracterización biogeográfica de la Faja Volcánica Transmexicana y análisis de los patrones de distribución de su mastofauna. Rev Mex Biodivers 83:258–272

García Díaz R, Cuevas Sánchez JA, Segura Ledesma S, Basurto Peña F (2015) Análisis panbiogeográfico de *Diospyros* spp. (Ebenaceae) en México. Rev Mex Cienc Agríc 6:187–200

García-Calderón NE, Krasilnikov P, Valeria Pérez MA, Torres Trejo E (2007) Suelos. In: Luna I, Morrone JJ, Espinosa D (eds) Biodiversidad de la Faja Volcánica Transmexicana. Las Prensas de Ciencias, UNAM, Mexico City, pp 73–98

García-Moreno J, Navarro-Sigüenza AT, Peterson T, Sánchez-González LA (2004) Genetic variation coincides with geographic structure in the common bush-tanager (*Chlorospingus ophthalmicus*) complex from Mexico. Mol Phylogenet Evol 33:186–196

García-Navarrete PG, Morrone JJ (2018) Testing the biogeographical regionalization of the Mexican transition zone based on the distribution of Curculionidae (Coleoptera). Zootaxa 4530:1–99

Goldman EA, Moore RT (1945) The biotic provinces of Mexico. J Mammal 26:347–360

Gómez-Ortiz Y, Domínguez-Vega H, Moreno CE (2017) Spatial variation of mammal richness, functional and phylogenetic diversity in the Mexican transition zone. Community Ecol 18:121–127

González-Ávila A, Contreras-Medina R, Espinosa D, Luna-Vega I (2017) Track analysis of the order Gomphales (Fungi: Basidiomycota) in Mexico. Phytotaxa 316:22–38

González-Elizondo MS, González-Elizondo M, Tena-Flores JA, Ruacho-González L, López-Enríquez IL (2012) Vegetación de la Sierra Madre Occidental, México: Una síntesis. Acta Bot Mex 100:351–403

González-Zamora A, Luna-Vega I, Villaseñor JL, Ruiz-Jiménez CA (2007) Distributional patterns and conservation of species of Asteraceae (asters etc.) endemic to eastern Mexico: a panbiogeographical approach. Syst Biodivers 5:135–144

Graham A (2004) A natural history of the New World: the ecology and evolution of plants in the Americas. The University of Chicago Press, Chicago and London

Gutiérrez-Ortega JS, Salinas-Rodríguez MM, Martínez JF, Molina-Freaner F, Pérez-Farrera MA, Vovides AP, Matsuki Y, Suyama Y, Ohsawa TA, Watano Y, Kajita T (2018) The phylogeography of the cycad genus *Dioon* (Zamiaceae) clarifies its Cenozoic expansion and diversification in the Mexican transition zone. Ann Bot 121:535–548

Gutiérrez-Velázquez AL, Acosta-Gutiérrez R (2004) Relaciones biogeográficas basadas en la distribución de Siphonaptera (Insecta). In: Luna I, Morrone JJ, Espinosa D (eds) Biodiversidad de la Sierra Madre Oriental. Las Prensas de Ciencias, UNAM, Mexico City, pp 393–416

Gutiérrez-Velázquez AL, Rojas-Soto O, Reyes-Castillo P, Halffter G (2013) The classic theory of Mexican transition zone revisited: the distributional congruence patterns of Passalidae (Coleoptera). Invert Syst 27:282–293

Halffter G (1962) Explicación preliminar de la distribución geográfica de los Scarabaeidae mexicanos. Acta Zool Mex 5:1–17
Halffter G (1964) Las regiones Neártica y Neotropical, desde el punto de vista de su entomofauna. An II Congr Latinoam Zool São Paulo 1962(1):51–61
Halffter G (1965) Algunas ideas acerca de la zoogeografía de América. Rev Soc Mex Hist Nat 26:1–16
Halffter G (1974) Eléments anciens de l'entomofaune neotropicale: Ses implications biogéographiques. Quaest Entomol 10:223–262
Halffter G (1976) Distribución de los insectos en la Zona de Transición Mexicana: Relaciones con la entomofauna de Norteamérica. Folia Entomol Mex 35:1–64
Halffter G (1978) Un nuevo patrón de dispersión en la zona de transición mexicana: El mesoamericano de montaña. Folia Entomol Mex 39-40:219–222
Halffter G (1987) Biogeography of the montane entomofauna of Mexico and Central America. Annu Rev Entomol 32:95–114
Halffter G, Morrone JJ (2017) An analytical review of Halffter's Mexican transition zone, and its relevance for evolutionary biogeography, ecology and biogeographical regionalization. Zootaxa 4226:1–46
Halffter G, Llorente-Bousquets J, Morrone JJ (2008) La perspectiva biogeográfica histórica. In: Sarukhán J (ed) Capital natural de México, Conocimiento actual de la biodiversidad, vol I. Conabio, Mexico City, pp 67–86
Halffter G, Zunino M, Moctezuma V, Sánchez-Huerta JL (2019) The integration of the distributional patterns in the Mexican transition zone: phyletic, paleogeographic and ecological factors of a case study. Zootaxa 4586:1–34
Harris D, Rogers DS, Sullivan J (2000) Phylogeography of *Peromyscus furvus* (Rodentia; Muridae) based on cytochrome b sequence data. Mol Ecol 9:2129–2135
Heilprin A (1887) The geographical and geological distribution of animals. International Scientific Series, New York and London
Hernández-Cerda ME, Carrasco-Anaya G (2004) Climatología. In: Luna I, Morrone JJ, Espinosa D (eds) Biodiversidad de la Sierra Madre Oriental. Las Prensas de Ciencias, UNAM, Mexico City, pp 63–108
Hernández-Cerda ME, Azpra-Romero E, Aguilar-Zamora V (2016) Condiciones climáticas de la Sierra Madre del Sur. In: Luna-Vega I, Espinosa D, Contreras-Medina R (eds) Biodiversidad de la Sierra Madre del Sur: Una síntesis preliminar. UNAM, Mexico City, pp 91–106
Huber O, Riina R (2003) Glosario ecológico de las Américas. Vol. 2. América del Sur: Países hispanoparlantes. UNESCO, Caracas
Katinas L, Crisci JV, Wagner WL, Hoch PC (2004) Geographical diversification of tribes Epilobiae, Gongylocarpae, and Onagreae (Onagraceae) in North America, based on parsimony analysis of endemicity and track compatibility analysis. Ann Missouri Bot Gard 91:159–185
Klassa B, Santos CMD (2015) Areas of endemism in the Neotropical region based on the geographical distribution of Tabanomorpha (Diptera: Brachycera). Zootaxa 4058:519–534
Kobelkowsky-Vidrio T, Ríos-Muñoz CA, Navarro-Sigüenza AG (2014) Biodiversity and biogeography of the avifauna of the Sierra Madre Occidental, Mexico. Biodivers Conserv 23:2087–2105
Kohlmann B, Halffter G (1988) Cladistic and biogeographical analysis of *Ateuchus* (Coleoptera: Scarabaeidae) of Mexico and the United States. Folia Entomol Mex 74:109–130
Kohlmann B, Halffter G (1990) Reconstruction of a specific example of insect invasion waves: the cladistic analysis of *Canthon* (Coleoptera: Scarabaeidae) and related genera in North America. Quaest Entomol 26:1–28
Lamas CJE, Nihei SS, Cunha AM, Couri MS (2014) Phylogeny and biogeography of *Heterostylum* (Diptera: Bombyliidae): evidence for an ancient Caribbean diversification model. Florida Entomol 97:952–966
León-Paniagua L, Morrone JJ (2009) Do the Oaxacan highlands represent a natural biotic unit? A cladistic biogeographical test based on vertebrate taxa. J Biogeogr 36:1939–1944

Léon-Paniagua L, García Trejo E, Arroyo-Cabrales J, Castañeda-Rico S (2004) Patrones biogeográficos de la mastofauna. In: Luna I, Morrone JJ, Espinosa D (eds) Biodiversidad de la Sierra Madre Oriental. Las Prensas de Ciencias, UNAM, Mexico City, pp 469–486

León-Paniagua L, Navarro AG, Hernández B, Morales JC (2007) Diversification of arboreal mice of genus *Habromys* (Rodentia: Cricetidae: Neotominae). Mol Phylogenet Evol 62:653–664

Liebherr JK (1991) A general area cladogram for montane Mexico based on distributions in the Platynine genera *Elliptoleus* and *Calathus* (Coleoptera: Carabidae). Proc Entomol Soc Washington 93:390–406

Liebherr JK (1994a) Biogeographic patterns of montane Mexican and central American Carabidae (Coleoptera). Can Entomol 126:841–860

Liebherr JK (1994b) Identification of New World *Agonum*, review of the Mexican fauna, and description of *Incagonum*, new genus, from South America (Coleoptera: Carabidae: Platynini). J N Y Entomol Soc 102:1–55

Llorente Bousquets J (1996) Biogeografía de artrópodos de México: ¿Hacia un nuevo enfoque? In: Llorente Bousquets J, Aldrete ANG, Soriano EG (eds) Biodiversidad, taxonomía y biogeografía de artrópodos de México: Hacia una síntesis de su conocimiento. Universidad Nacional Autónoma de México, Mexico City, pp 41–56

Luna I, Morrone JJ, Espinosa D (eds) (2004) Biodiversidad de la Sierra Madre Oriental. Las Prensas de Ciencias, UNAM, Mexico City

Luna I, Morrone JJ, Espinosa D (eds) (2007) Biodiversidad de la Faja Volcánica Transmexicana. FES Zaragoza, UNAM, Mexico City

Luna Vega I, Alcántara Ayala O, Espinosa Organista D, Morrone JJ (1999) Historical relationships of the Mexican cloud forests: a preliminary vicariance model applying parsimony analysis of endemicity to vascular plant taxa. J Biogeogr 26:1299–1305

Luna-Vega I, Espinosa D, Contreras-Medina R (eds) (2016) Biodiversidad de la Sierra Madre del Sur: Una síntesis preliminar. UNAM, Mexico City

Lydekker BA (1896) A geographical history of mammals. Cambridge University Press, Cambridge

Mariño-Pérez R, Brailovsky H, Morrone JJ (2007) Análisis panbiogeográfico de las especies mexicanas de *Pselliopus* Bergroth (Hemiptera: Heteroptera: Reduviidae: Harpactorinae). Acta Zool Mex (n. s.) 23:77–88

Márquez J, Morrone JJ (2003) Análisis panbiogeográfico de las especies de *Heterolinus* y *Homalolinus* (Coleoptera: Staphylinidae: Xantholinini). Acta Zool Mex (n. s.) 90:15–25

Márquez J, Morrone JJ (2004) Relaciones biogeográficas basadas en la distribución de Coleoptera (Insecta). In: Luna I, Morrone JJ, Espinosa D (eds) Biodiversidad de la Sierra Madre Oriental. Las Prensas de Ciencias, UNAM, Mexico City, pp 375–392

Marshall CJ, Liebherr JK (2000) Cladistic biogeography of the Mexican transition zone. J Biogeogr 27:203–216

Martínez-Aquino A, Aguilar-Aguilar R, Santa Anna del Conde Juárez H, Contreras-Medina R (2007) Empleo de herramientas panbiogeográficas para detectar áreas para conservar: Un ejemplo con taxones dulceacuícolas. In: Luna I, Morrone JJ, Espinosa D (eds) Biodiversidad de la Faja Volcánica Transmexicana. Las Prensas de Ciencias, UNAM, Mexico City, pp 449–460

Martínez-Icó M, Cetzal-Ix W, Noguera-Savelli E, Hernández-Juárez R (2015) Flora vascular de la comunidad de Bazom, Los Altos de Chiapas, México. Bot Sci 93:53–72

Matthew WD (1915) Climate and evolution. Ann N Y Acad Sci 24:171–318

Mayr E (1946) History of the north American bird fauna. Wilson Bull 58:3–41

Meave J, Ibarra-Manrique G, Larson-Guerra J (2016) Vegetación: Panorama histórico, rasgos generales y patrones de pérdida. In: Moncaya M, López-López A (eds) Geografía de México: Una reflexión espacial contemporánea. UNAM, Mexico City, pp 216–234

Mercado-Salas NF, Pozo C, Morrone JJ, Suárez-Morales E (2012) Distribution patterns of the American species of the freshwater genus *Eucyclops* (Copepoda: Cyclopoida). J Crust Biol 32:457–464

Miguez-Gutiérrez A, Castillo J, Márquez J, Goyenechea I (2013) Biogeografía cladística de la zona de transición Mexicana con base en un análisis de árboles reconciliados. Rev Mex Biodivers 84:215–224
Miller RR (1966) Geographical distribution of Central American freshwater fishes. Copeia 4:773–802
Moore RT (1945) The transverse volcanic biotic province of Central Mexico and its relationship to adjacent provinces. Trans San Diego Soc Nat Hist 10:217–236
Morón MA (1996) Scarabaeidae (Coleoptera). In: Llorente Bousquets J, Aldrete ANG, Soriano EG (eds) Biodiversidad, taxonomía y biogeografía de artrópodos de México: Hacia una síntesis de su conocimiento. Universidad Nacional Autónoma de México, Mexico City, pp 309–328
Morrone JJ (2001) Biogeografía de América Latina y el Caribe. Vol. 3. Manuales & Tesis SEA. Sociedad Entomológica Aragonesa, Zaragoza
Morrone JJ (2004) Panbiogeografía, componentes bióticos y zonas de transición. Rev Bras Entomol 48:149–162
Morrone JJ (2005) Hacia una síntesis biogeográfica de México. Rev Mex Biodivers 76:207–252
Morrone JJ (2006) Biogeographic areas and transition zones of Latin America and the Caribbean Islands based on panbiogeographic and cladistic analyses of the entomofauna. Annu Rev Entomol 51:467–494
Morrone JJ (2010) Fundamental biogeographic patterns across the Mexican transition zone: an evolutionary approach. Ecography 33:355–361
Morrone JJ (2014a) Biogeographical regionalisation of the Neotropical region. Zootaxa 3782:1–110
Morrone JJ (2014b) Cladistic biogeography of the Neotropical region: identifying the main events in the diversification of the terrestrial biota. Cladistics 30:202–214
Morrone JJ (2017a) Neotropical biogeography: regionalization and evolution. CRC Press, Boca Raton, FL
Morrone JJ (2017b) Biogeographic regionalization of the Sierra Madre del Sur province, Mexico. Rev Mex Biodivers 81:561–568
Morrone JJ (2018) The spectre of biogeographic regionalization. J Biogeogr 45:282–288
Morrone JJ (2019) Regionalización biogeográfica y evolución biótica de México: Encrucijada de la biodiversidad del Nuevo Mundo. Rev Mex Biodivers 90:1–68
Morrone JJ, Gutiérrez A (2005) Do fleas (Insecta: Siphonaptera) parallel their mammal host diversification in the Mexican transition zone? J Biogeogr 32:1315–1325
Morrone JJ, Llorente Bousquets J (2006) Conclusiones. In: Morrone JJ, Llorente Bousquets J (eds) Componentes bióticos principales de la entomofauna mexicana, Vol. II, Las Prensas de Ciencias. UNAM, Mexico City, pp 1011–1025
Morrone JJ, Márquez J (2001) Halffter's Mexican transition zone, beetle generalised tracks, and geographical homology. J Biogeogr 28:635–650
Morrone JJ, Márquez J (2003) Aproximación a un Atlas Biogeográfico de México: Componentes bióticos principales y provincias biogeográficas. In: Morrone JJ, Llorente J (eds) Una perspectiva latinoamericana de la biogeografía. Las Prensas de Ciencias, UNAM, Mexico City, pp 217–220
Morrone JJ, Márquez J (2008) Biodiversity of Mexican arthropods (Arachnida and Hexapoda): a biogeographical puzzle. Acta Zool Mex (n. s.) 24:15–41
Morrone JJ, Espinosa D, Aguilar Zúñiga C, Llorente Bousquets J (1999) Preliminary classification of the Mexican biogeographic provinces: a parsimony analysis of endemicity based on plant, insect, and bird taxa. Southwest Nat 44:507–514
Morrone JJ, Espinosa D, Llorente J (2002) Mexican biogeographic provinces: preliminary scheme, general characterizations, and synonymies. Acta Zool Mex (n. s.) 85:83–108
Morrone JJ, Escalante T, Rodríguez-Tapia G (2017) Mexican biogeographic provinces: map and shapefiles. Zootaxa 4277:277–279
Müller P (1973) The dispersal centers of terrestrial vertebrates in the Neotropical realm: a study in the evolution of the Neotropical biota and its native landscapes. Junk, The Hague

Müller P (1986) Biogeography. Harper & Row, New York

Munguía-Lino G, Escalante T, Morrone JJ, Rodríguez A (2016) Areas of endemism of the north American species of Tigridieae (Iridaceae). Aust Syst Bot 29:142–156

Navarro AG, Graza-Torres HA, López de Aquino S, Rojas-Soto OR, Sánchez-González LA (2004) Patrones biogeográficos de la avifauna. In: Luna I, Morrone JJ, Espinosa D (eds) Biodiversidad de la Sierra Madre Oriental. Las Prensas de Ciencias, UNAM, Mexico City, pp 439–467

Navarro-Sigüenza AG, Lira-Noriega A, Peterson AT, Oliveras de Ita A, Gordillo-Martínez A (2007) Diversidad, endemismo y conservación de las aves. In: Luna I, Morrone JJ, Espinosa D (eds) Biodiversidad de la Faja Volcánica Transmexicana. Las Prensas de Ciencias, UNAM, Mexico City, pp 462–483

Navarro-Sigüenza AG, Blancas-Calva E, Almazán-Núñez RC, Hernández-Baños BE, García-Trejo EA, Peterson AT (2016) Diversidad y endemismo de las aves de la Sierra Madre del Sur. In: Luna-Vega I, Espinosa D, Contreras-Medina R (eds) Biodiversidad de la Sierra Madre del Sur: Una síntesis preliminar. UNAM, Mexico City, pp 381–411

Nolasco-Soto J, González-Astorga J, Espinosa de los Monteros A, Galente-Patiño E, Favila ME (2017) Phylogeographic structure of *Canthon cyanellus* (Coleoptera: Scarabaeidae), a Neotropical dung beetle in the Mexican transition zone: insights on its origin and the impacts of Pleistocene fluctuations on population dynamics. Mol Phylogenet Evol 109:180–190

Ortega J, Arita HT (1998) Neotropical-Nearctic limits in middle America as determined by distributions of bats. J Mammal 79:772–781

Perry JP, Graham A, Richardson DM (2000) The history of pines in Mexico and Central America. In: Richardson DM (ed) Ecology and biogeography of *Pinus*. Cambridge University Press, Cambridge, UK, pp 137–149

Pinilla-Buitrago GE, Escalante T, Gutiérrez-Velázquez A, Reyes-Castillo P, Rojas-Soto OR (2018) Areas of endemism persist through time: a paleoclimatic analysis in the Mexican transition zone. J Biogeogr 45:952–961

Porzecanski AL, Cracraft J (2005) Cladistic analysis of distributions and endemism (CADE): using raw distributions of birds to unravel the biogeography of the south American aridlands. J Biogeogr 32:261–275

Puga-Jiménez AL, Andrés-Hernández AR, Carrillo-Ruiz H, Espinosa-Organista D, Rivas-Arancibia SP (2013) Patrones de distribución del género *Zanthoxylum* L. (Rutaceae) en México. Rev Mex Biodivers 84:1179–1188

Quijano-Abril MA, Callejas-Posada R, Miranda-Esquivel DR (2006) Areas of endemism and distribution patterns for Neotropical Piper species (Piperaceae). J Biogeogr 33:1266–1278

Ramírez-Pulido J, Castro-Campillo A (1990) Regionalización mastofaunística (mamíferos). Mapa IV. 8. 8. A. In: Atlas Nacional de México. vol III, Instituto de Geografía, UNAM, Mexico City, map

Ramírez-Pulido J, Castro-Campillo A, Salame-Méndez A (2005) Relación de algunas especies del género *Reithrodontomys* (Rodentia: Muridae) en la colección de mamíferos de la UAM. In: Sánchez-Cordero V, Medellín RA (eds) Contribuciones mastozoológicas en homenaje a Bernardo Villa. Instituto de Biología, UNAM and Conabio, Mexico City, pp 399–422

Ríos-Muñoz CA (2013) ¿Es posible reconocer una unidad biótica entre América del Norte y del Sur? Rev Mex Biodivers 84:1022–1030

Rodríguez A, Castro-Castro A, Vargas-Amado G, Vargas-Ponce O, Zamora-Tavares P, González-Gallegos J, Carrillo-Reyes P, Anguiano-Constante M, Carrasco-Ortiz M, García-Martínez M, Gutiérrez-Rodríguez B, Aragón-Parada J, Valdés-Ibarra C, Munguía-Lino G (2018) Richness, geographic distribution patterns, and areas of endemism of selected angiosperm groups in Mexico. J Syst Evol 56:537–549

Rodríguez-Gómez F, Ornelas JF (2015) At the passing gate: past introgression in the process of species formation between *Amazilia violiceps* and *A. viridifrons* hummingbirds along the Mexican transition zone. J Biogeogr 42:1305–1318

Ruiz-Sánchez E, Specht CD (2013) Influence of the geological history of the trans-Mexican Volcanic Belt on the diversification of *Nolina parviflora* (Asparagaceae: Nolinoideae). J Biogeogr 40:1336–1347

Ryan RM (1963) The biotic provinces of Central America. Acta Zool Mex 6:1–55

Rzedowski J (1963) Contribuciones a la fitogeografía florística e histórica de México. I. Algunas consideraciones acerca del elemento endémico en la flora mexicana. Bol Soc Bot México 27:52–65

Rzedowski J (1978) La vegetación de México. Editorial Limusa, Mexico City

Rzedowski J (1991) Diversidad y orígenes de la flora fanerogámica de México. Acta Bot Mex 14:3–21

Rzedowski J, Reyna-Trujillo T (1990) Tópicos biogeográficos. Mapa IV. 8. 3. In: Atlas Nacional de México, vol. III. Instituto de Geografía, UNAM, Mexico City

Salinas-Rodríguez MM, Estrada-Castillón E, Villarreal-Quintanilla JA (2017) Endemic vascular plants of the Sierra Madre Oriental, Mexico. Phytotaxa 328:1–52

Sánchez-González LA, Morrone JJ, Navarro-Sigüenza A (2008) Distributional patterns of the Neotropical humid montane forest avifaunas. Biol J Linn Soc 94:175–194

Sanginés-Franco C, Luna-Vega I, Alcántara Ayala O, Contreras-Medina R (2011) Distributional patterns and biogeographic analysis of ferns in the Sierra Madre Oriental, Mexico. Am Fern J 101:81–104

Santa Anna del Conde Juárez H, Contreras-Medina R, Luna-Vega I (2009) Biogeographic analysis of endemic cacti of the Sierra Madre Oriental, Mexico. Biol J Linn Soc 97:373–389

Santa María-Díaz A (2016) Aspectos geológicos de la Sierra Madre del Sur. In: Luna-Vega I, Espinosa D, Contreras-Medina R (eds) Biodiversidad de la Sierra Madre del Sur: Una síntesis preliminar. UNAM, Mexico City, pp 39–65

Santiago-Alvarado M (2019) Evaluación de la validez de las provincias de la Zona de Transición Mexicana. M.Sc. dissertation, FES Zaragoza, UNAM, Mexico City

Santiago-Alvarado M, Montaño-Arias G, Espinosa D (2016) Áreas de endemismo de la Sierra Madre del Sur. In: Luna-Vega I, Espinosa D, Contreras-Medina R (eds) Biodiversidad de la Sierra Madre del Sur: Una síntesis preliminar. UNAM, Mexico City, pp 431–448

Savage JM (1966) The origins and history of Central America herpetofauna. Copeia 4:719–766

Sclater WL, Sclater PL (1899) The geography of mammals. Kegan Paul, Trench, Trübner & Co., London

Smith H (1941) Las provincias bióticas de México, según la distribución geográfica de las lagartijas del género *Sceloporus*. An Esc Nac Cienc Biol 2:103–110

Stuart LC (1964) Fauna of Middle America. In: West RC (ed) Handbook of Middle American Indians, vol 1. University of Texas Press, Austin, pp 316–363

Suárez-Mota M, Téllez-Valdés O, Lira-Saade R, Villaseñor JL (2013) Una regionalización de la Faja Volcánica Transmexicana con base en su riqueza florística. Bot Sci 91:93–105

Tiemann-Boege I, Kilpatrick CW, Schmidly DJ, Bradley RD (2000) Molecular phylogenetics of the *Peromyscus boylii* species group (Rodentia: Muridae) based on mitochondrial cytochrome b sequences. Mol Phylogenet Evol 16:366–378

Torres Miranda A, Luna I (2007) Hacia una síntesis panbiogeográfica. In: Luna I, Morrone JJ, Espinosa D (eds) Biodiversidad de la Faja Volcánica Transmexicana. Las Prensas de Ciencias, UNAM, Mexico City, pp 503–514

Torres Miranda A, Luna Vega I (2006) Análisis de trazos para establecer áreas de conservación en la Faja Volcánica Transmexicana. Interciencia 31:849–855

Townsend JH (2014) Characterizing the Chortís block biogeographic province: geological, physiographic, and ecological associations and herpetofaunal diversity. Mesoamer Herpetol 1:204–251

Valero A, Schipper J, Allnutt T, Burdette C (2019a) Southern North America: Western Mexico into the southwestern United States. In: Terrestrial ecoregions of Latin America and the Caribbean, World Wildlife Fund. WWF NT1310. https://www.worldwildlife.org/ecoregions/na1310. Accessed 9 Aug 2019

Valero A, Schipper J, Allnutt T, Burdette C (2019b) Southern North America: Eastern Mexico into southwestern United States. In: Terrestrial ecoregions of Latin America and the Caribbean, World Wildlife Fund. WWF NT0303. https://www.worldwildlife.org/ecoregions/na0303. Accessed 9 Aug 2019

Valero A, Schipper J, Allnutt T, Burdette C (2019c) Southern North America: Southern Mexico. In: Terrestrial ecoregions of Latin America and the Caribbean, World Wildlife Fund. WWF NT0310. https://www.worldwildlife.org/ecoregions/nt0310. Accessed 9 Aug 2019

Velasco de León P, Arellano Gil J, Silva-Pineda A, Yussim Guarneros S (2007) Aspectos geológicos y paleontológicos. In: Luna I, Morrone JJ, Espinosa D (eds) Biodiversidad de la Faja Volcánica Transmexicana. Las Prensas de Ciencias, UNAM, Mexico City, pp 25–38

Velasco de León P, Arellano-Gil J, Ortiz-Martínez EL, Lozano-Carmona DE, Domínguez-Trejo I, Canales-García I, Carbot-Chanona G (2016) Paleontología y geología de la Sierra Madre del Sur. In: Luna-Vega I, Espinosa D, Contreras-Medina R (eds) Biodiversidad de la Sierra Madre del Sur: Una síntesis preliminar. UNAM, Mexico City, pp 67–90

Villaseñor JL, Ortiz E (2013) Biodiversidad de las plantas con flores (división Magnoliophyta) en México. Rev Mex Biodivers 85:S134–S142

Wallace AR (1876) The geographical distribution of animals, vol I and II. Harper & Brothers, New York

West RC (1964) The natural regions of Middle America. In: West RC (ed) Handbook of Middle American Indians, vol 1. University of Texas Press, Austin, pp 363–383

Yanahan AD, Moore W (2019) Impacts of 21st-century climate change on montane habitat in the Madrean Sky Island Archipelago. Divers Distrib 25:1625–1638

Zamora-Tavares MP, Martínez M, Magallón S, Guzmán-Dávalos L, Vargas-Ponce O (2016) *Physalis* and physaloids: a recent and complex evolutionary history. Mol Phylogenet Evol 100:41–50

Zunino M, Halffter G (1988) Análisis taxonómico, ecológico y biogeográfico de un grupo americano de *Onthophagus* (Coleoptera: Scarabaeidae). Monogr Mus Reg Sci Nat 9:1–211

Chapter 5
The Biotic Assembly of the Mexican Transition Zone

It is not that we propose a theory and Nature shouts NO; rather, we propose a maze of theories, and Nature may shout INCONSISTENT.

Imre Lakatos (1980), The Methodology of Scientific Research Programmes

Abstract The biota of the Mexican Transition Zone was assembled through the successive dispersal of four cenocrons from North and South America and their incorporation to the Paleoamerican biota, which was the original North American (Holarctic) biota that extended in Mexico in the Jurassic-Cretaceous. The Mexican Plateau cenocron dispersed to southern North America from South America (Gondwana) in the Late Cretaceous-Paleocene. The Mountain Mesoamerican cenocron dispersed from South America to the mountain forests of Central America and southern Mexico in the Oligocene-Miocene and then northward in the Pliocene. The Nearctic cenocron dispersed from northern North America to the mountains of the Mexican Transition Zone in the Miocene-Pliocene. Finally, in the Pleistocene the Typical Neotropical cenocron dispersed from the Neotropical region, being represented by genera widespread in South America. Based on the successive assembly of these cenocrons, three horobiotas are distinguished: Paleogene horobiota (original Paleoamerican biota plus Mexican Plateau cenocron), Neogene horobiota (Paleoamerican biota plus Mexican Plateau, Mountain Mesoamerican, and Nearctic cenocrons), and Quaternary horobiota (Paleoamerican biota plus Mexican Plateau, Mountain Mesoamerican, Nearctic and Typical Neotropical cenocrons)

5.1 Biotic Assembly in a Biogeographic Transition Zone

The study of biotic assembly in transition zones is still in its infancy (Ferro and Morrone 2014). We do not have yet a well-formulated theory like those developed for community assembly on islands (e.g., Whittaker and Fernández-Palacios 2007),

J. J. Morrone, *The Mexican Transition Zone*, https://doi.org/10.1007/978-3-030-47917-6_5

and we lack some sort of assembly rules like those developed by Diamond (1975). Weeks et al. (2016) recently proposed a protocol for historical assembly analysis, which may be adapted for the analysis of transition zones. Their protocol is focused on reconstructing the evolutionary history of individual lineages, dating their divergence times, discovering the underlying processes guiding their assembly, and finally matching the temporal patterns of biotic and Earth history. I apply Weeks et al.'s (2016) protocol herein, with the incorporation of the concepts of cenocrons and horobiotas, to the Mexican Transition Zone, and use the original contributions of Halffter (e.g., 1964, 1976, 1978, 1987) as the bases of my analysis.

The Mexican Transition Zone owes much of its characteristics to its physiography. Mountains, especially those placed in the tropics, have proven to exhibit extraordinary species richness and endemism, which have been related to several factors (Rahbek et al. 2019). Mountains situated in the tropics encompass different climatic types, which are in close proximity with one another in a relatively small space when compared to lowlands. Short-term climatic oscillations (e.g., diurnal variation of temperature, seasonality, etc.) and longer-term climatic changes are unique in mountains, leading to periodic reconfigurations of the connection among mountain habitats. Physical parameters drop markedly with increasing elevation, leading to specific adaptations to high-mountain conditions. The geophysical variation of mountains (e.g., different slopes) interacts with the spatial and temporal variation, producing many rare climate types, restricted to mountains that exhibit exceptional combinations of climatic factors. All these features interact to generate particular conditions for the biotic richness of mountains (Rahbek et al. 2019).

In this chapter, I characterize first the original Paleoamerican biota and then the different cenocrons that assembled to it. Based on the successive assembly of the cenocrons, three different horobiotas are recognized. The phylogenetic relationships and the geographic distribution of the lineages assigned to the different cenocrons and the timing of their divergence are used to reconstruct the dispersal and vicariance events that led to their assembly. Finally, a geobiotic scenario encompassing tectonic and geomorphological studies articulates the assembly of cenocrons in space and time.

5.2 The Paleoamerican Distributional Pattern: The Original Biota

The Paleoamerican distributional pattern (Halffter 1964, 1987; Zunino and Halffter 1988) corresponds to the original North American biota. It is comprised of septentrional (Holarctic) taxa that were already present in North America since the Jurassic-Cretaceous or that dispersed to it from Eurasia in the Early Cenozoic. Paleoamerican clades have species in the Mexican Plateau, the mountains of the Mexican Transition Zone, and they may also have species in the tropical lowlands

(Halffter and Morrone 2017). Most Paleoamerican taxa are almost cosmopolitan, being common in the tropics and temperate areas of the world.

The Paleoamerican biota comprises the oldest taxa of the Mexican Transition Zone (Halffter 1987, 2017; Halffter and Morrone 2017). They evolved originally in Eurasia or Africa and extended to North America in the Jurassic-Cretaceous, before the formation of the Mexican Plateau and the expansion of the deserts of western North America. Species belonging to the same Paleoamerican genus may exhibit adaptations to very different climatic conditions. Their dispersal to the Mexican Transition Zone has been partially synchronic with the dispersal of old Neotropical lineages from South America, namely, the Mexican Plateau pattern (Halffter 1976).

Plant taxa belonging to this Paleoamerican biota have been characterized by paleobotanists as the boreotropical flora. It corresponds to the flora that developed in the Northern Hemisphere during the Early Paleogene (Lavin and Luckow 1993). The plant genus *Dioon* (Zamiaceae) is an example of a Paleoamerican taxon (Gutiérrez-Ortega et al. 2018). It has a relictual distribution in different geographic areas of the Mexican Transition Zone and represents an excellent model for phylogeographic analyses. *Dioon* has an estimated age of 56 m. y. and has diversified in areas where different ecological and climatic factors might have promoted speciation (Gutiérrez-Ortega et al. 2018).

Paleoamerican insect taxa include the genera *Onthophagus* and *Copris* (Coleoptera: Scarabaeidae) and *Bombus* (Hymenoptera: Apidae). *Onthophagus*, a taxon that apparently evolved originally in tropical Africa, is practically cosmopolitan and is distributed from sea level to high altitude areas. In the Americas, it has at least 158 native species (from a world total of 1200 species), most of them distributed in the Mexican Transition Zone (Pulido-Herrera and Zunino 2007). *Phloexena* and *Eripus* (Coleoptera: Carabidae) are other examples of Paleoamerican taxa (Liebherr 1994).

The lizard genus *Sceloporus* (Squamata: Phrynosomatidae) is another Paleoamerican taxon. It shows a remarkable diversification in the United States, Mexico, and Central America, with different species groups exhibiting characteristic distributional areas (Sites et al. 1992). *Sceloporus* was originally hypothesized by Savage (1982) to represent a Young Northern element (=Typical Neotropical cenocron). Sites et al. (1992), however, considered that Phrynosomatidae evolved in the Cretaceous, so *Sceloporus* may be considered to belong to the Paleoamerican biota (=Old Northern element of Savage 1982).

The analysis of some Paleoamerican taxa, e.g., the *Onthophagus chevrolati* species group (Coleoptera: Scarabaeidae), has shown that some taxa assigned to this distributional pattern do not have common ecological features, because their ancient presence in the Mexican Transition Zone and their adaptive plasticity have allowed them to diversify ecologically (Zunino and Halffter 1988). Thus, it has been considered better to split the Paleoamerican pattern into six different subpatterns or varieties (Lobo and Halffter 2000; Halffter et al. 2008, 2019; Halffter 2017; Halffter and Morrone 2017; Moctezuma and Halffter 2019):

Relict Paleoamerican subpattern: it corresponds to species assigned to genera or other supraspecific taxa with wide geographical and ecological distribution in the Old World. They are represented in the Mexican Transition Zone by endemic species with very restricted distributions. A blind weevil genus (Coleoptera: Curculionidae) that corresponds to this subpattern is being currently described (Manuel Barrios-Izás pers. comm.).

Mountain Paleoamerican subpattern: it corresponds to lineages that have colonized successfully the mountains of the Mexican Transition Zone and, to a lesser extent, those of Central America. They have undergone significant speciation in the Mexican Transition Zone, driven by vicariance. One example of this subpattern is the *Onthophagus hippopotamus* lineage of the *O. chevrolati* species group (Coleoptera: Scarabaeidae), analyzed recently by Halffter et al. (2019). The current distribution of the *O. hippopotamus* species group is shown in the map of North America (Fig. 5.1), including *O. coproides* from the northern Sierra Madre Occidental and south-eastern United States. Another examples are the Scarabaeid genera *Ceratotrupes*, *Cnemotrupes*, *Geohowdenius*, *Onthotrupes*, and *Peltotrupes* that might have evolved in Mexico in the Miocene-Pliocene (Halffter and Morrone 2017).

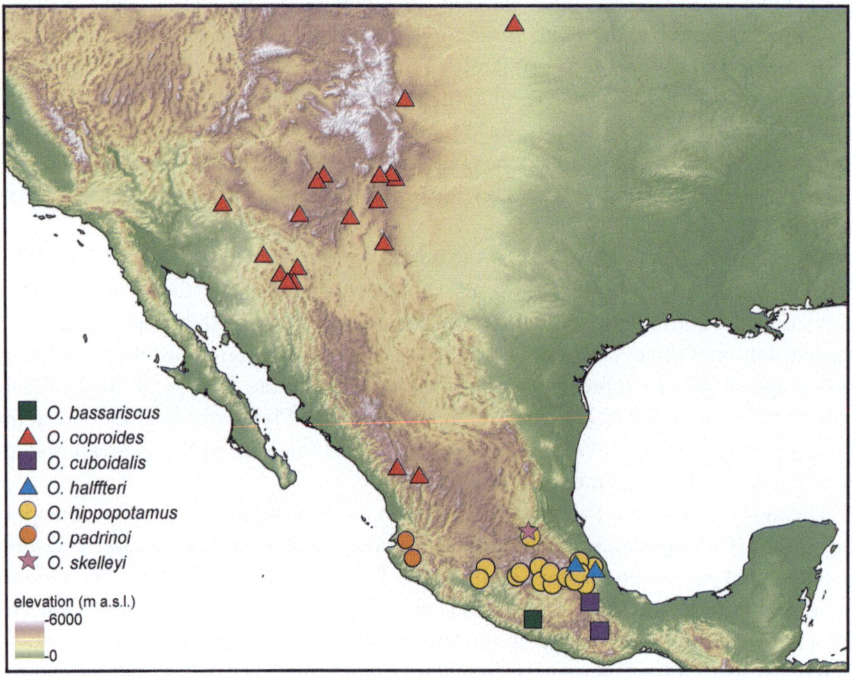

Fig. 5.1 Mountain Paleoamerican distributional pattern: *Onthophagus hippopotamus* species complex (Coleoptera: Scarabaeidae) (modified from Halffter et al. 2019)

Mesoamerican Paleoamerican subpattern: it corresponds to species that are found in tropical lowlands at moderate altitudes. These species have distributions that may be very similar to those of the Neotropical pattern, but their affinities are with Old World taxa. Examples of this subpattern are the *Onthophagus dicranius* species group and the *Copris remotus* species complex (Coleoptera: Scarabaeidae; Halffter et al. 2019).

Mexican Plateau Paleoamerican subpattern: it corresponds to Paleoamerican species distributed in the Mexican Plateau. Although Halffter (1976) and Zunino and Halffter (1988) highlighted the existence of such taxa, they did not establish formally a subpattern. It was formalized by Halffter et al. (2008): "In terms of its phylogenetic-biogeographic-historical composition, the Plateau is an excellent example of the complexity of the Mexican Transition Zone. Although Neotropical taxa dominate, there is a Paleoamerican component represented by genera of northern origins: *Onthophagus* and *Copris*" (Halffter et al. 2008, p. 297). Arriaga et al. (2012) and Barragán et al. (2014) also referred to the Mexican Plateau Paleoamerican subpattern, considering that it is clearly distinguished from the Mountain Mesoamerican and Tropical Paleoamerican subpatterns by their origin and phylogenetic affinities.

Tropical Paleoamerican subpattern: it corresponds to species that follow tropical conditions characteristic of the lowlands and moderate altitudes. Their distributions are quite similar to those of the Typical Neotropical pattern in their boundaries but are completely different from the phylogenetic viewpoint. One example of this subpattern is the *Onthophagus clypeatus* species group (Coleoptera: Scarabaeidae) (Halffter and Morrone 2017).

Baja Californian Paleoamerican subpattern: it corresponds to taxa with Holarctic affinities that are endemic to the southern Baja California Peninsula, as a result of a Miocene terrestrial dispersal from mainland Mexico and subsequent vicariance. One example of this subpattern is *Onthophagus bajacalifornianus* (Coleoptera: Scarabaeidae; Moctezuma and Halffter 2019).

Halffter et al. (2019) provided an interesting analysis of the diversification of a Paleoamerican taxon in the Mexican Transition Zone, the *Onthophagus chevrolati* species group (Coleoptera: Scarabaeidae). Based on a phylogenetic hypothesis (Fig. 5.2), the authors reconstructed the main events in its diversification. The ancestor of the *O. hippopotamus* species complex dispersed through the Transmexican Volcanic Belt before it attained its current conformation, and its geomorphological evolution produced vicariance events that induced allopatric speciation in the group. The *O. brevifrons* species complex is found in southeastern United States and northeastern Mexico. The species of both the *O. hippopotamus* and *O. brevifrons* species complexes are associated with rodent nests of Geomyidae and *Neotoma* (Cricetidae), respectively.

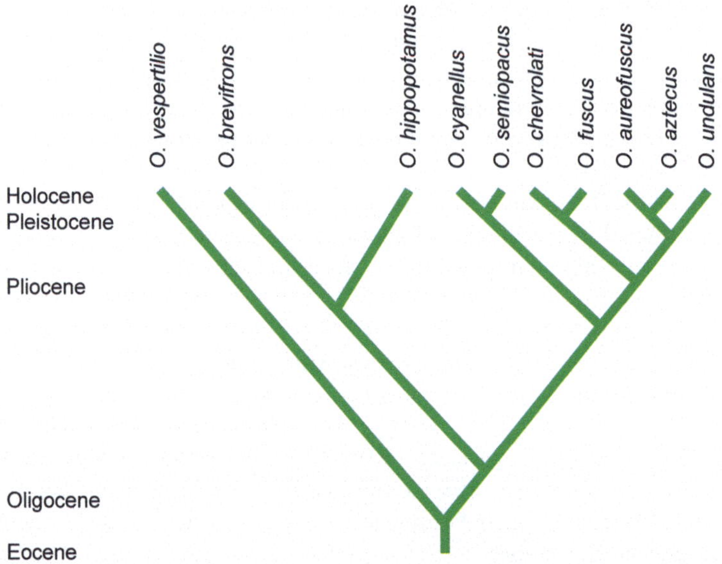

Fig. 5.2 Phylogenetic relationships of the species complexes of the *Onthophagus chevrolati* species group (Coleoptera: Scarabaeidae) (modified from Halffter et al. 2019)

5.3 The Mexican Plateau Cenocron: Old South American/Gondwanan Immigrants

The Mexican Plateau cenocron corresponds to South American taxa that dispersed into the Mexican Plateau in the Late Cretaceous-Paleocene (Halffter 1962, 1987; Halffter and Morrone 2017). The Mexican Plateau in the past represented geographically a "peninsula of the Nearctic region" (Halffter 1962). Species are different to those of the Typical Neotropical distributional pattern and have dispersed north of Mexico, originating Neotropical species in the southern United States.

Some plant taxa are characteristic of the Mexican Plateau cenocron. One example is the family Cactaceae, which evolved in northern South America in the Late Cretaceous. They might have dispersed to North America by island hopping along the Proto-Antilles (Yetman 2007). Another example is *Prosopis* (Fabaceae), which is distributed in the Americas, northern Africa, and India (Rzedowski 1988; Catalano et al. 2008). Diversification of the species of *Prosopis* both in North and South America seems to be correlated with the spread of arid environments that started in the Late Miocene and continued during the Pliocene (Catalano et al. 2008).

A typical insect taxon belonging to the Mexican Plateau cenocron is the *Canthon humectus* species group (Coleoptera: Scarabaeidae). This group (Fig. 5.3) consists of four species, one of which (*C. humectus*) comprises six subspecies, which reflects its extraordinary diversification (Halffter and Morrone 2017). In the expansion of the *Canthon humectus* species group in the western part of the Mexican Plateau,

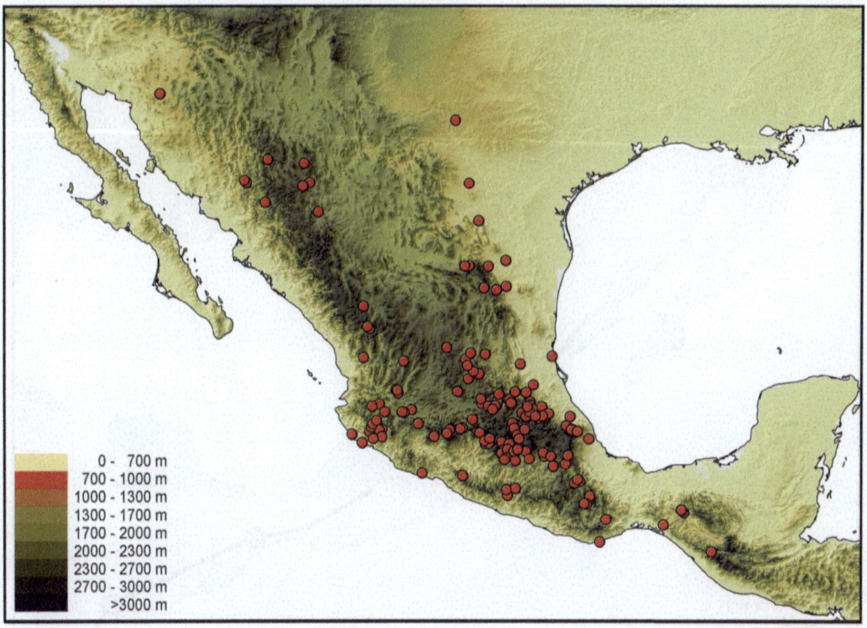

Fig. 5.3 Mexican Plateau distributional pattern: *Canthon humectus* species group (Coleoptera: Scarabaeidae) (modified from Halffter and Morrone 2017)

several species and subspecies are found in mountains, and one of them even reaches the Pacific coast. Halffter (1987) considered that the species of Scarabaeidae assigned to this cenocron constitute more than 80% of the species found in the Mexican Plateau. This abundance of Neotropical insect species contrasts clearly with the vertebrate fauna of the Mexican Plateau, which is characteristically Nearctic. One example of an old Neotropical dispersal to the Mexican Transition Zone discussed by Halffter and Morrone (2017) corresponds to the *Phanaeus quadridens* and *P. triangularis* species groups (Coleoptera: Scarabaeidae), whereas the *P. vindex* species group, closely related to the latter, is ranged in southern and southeastern United States. Another example is the weevil genus *Phacepholis* (Coleoptera: Curculionidae), analyzed by Rosas et al. (2011a, b). Based on the Mexican and Central American species of *Phacepholis*, *Pantomorus*, and *Naupactus* that belong to the Mexican Plateau and Mountain Mesoamerican cenocrons, Rosas et al. (2011b) found six generalized tracks and two nodes (Fig. 5.4).

The fish family Poeciliidae (Cyprinodontiformes) is an example of a Gondwanan immigrant in the Mexican Transition Zone. They represent a widely studied taxon (e.g., Rosen 1960, 1979; Parenti 1981). Hrbek et al. (2007) hypothesized that they dispersed from South America to North America in the Late Cretaceous (ca. 68 m.y.a.). Reznick et al. (2017) obtained a timetree calibrated with numerous fossils, concluding that Poeciliidae evolved in South America and then dispersed to North America, representing Central American taxa the result of a posterior dispersal event.

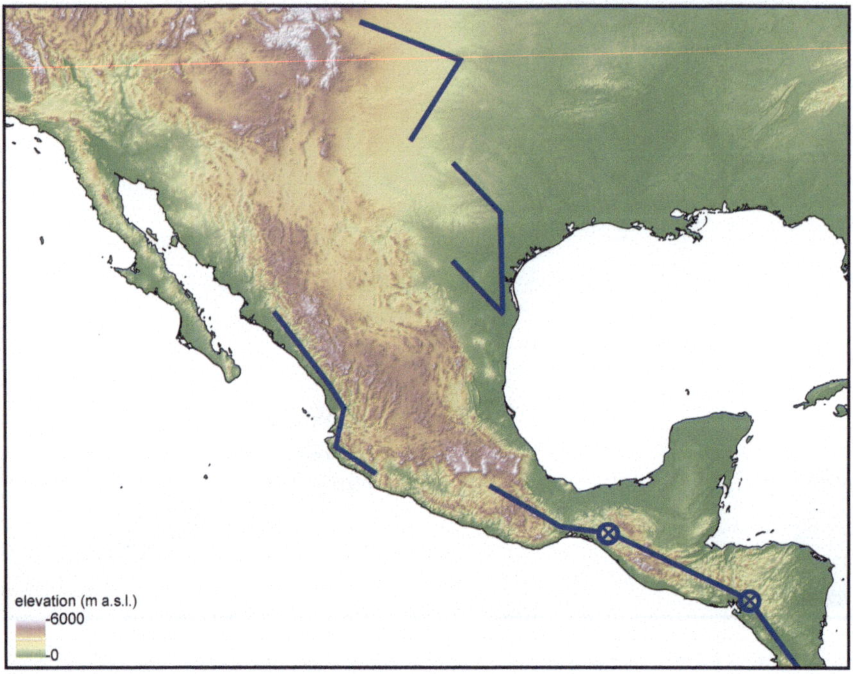

Fig. 5.4 Generalized tracks (lines) and nodes (circles enclosing an x) found by Rosas et al. (2011b). Generalized tracks represent biotas and node complex areas, where different biotas overlap

5.4 The Mountain Mesoamerican Cenocron: Central (and South) American Immigrants

The Mountain Mesoamerican cenocron was described originally by Halffter (1978) to accommodate some species of the mountain forests of Central America and southern Mexico. These taxa evolved in the Central American Nucleus, which corresponds to the highlands of Chiapas, Guatemala, Honduras, El Salvador, and Nicaragua north of Lake Nicaragua, in the Oligocene-Miocene, from South American ancestors, and dispersed northward in the Pliocene. The Central American Nucleus, which corresponds to the Chiapas Highlands biogeographic province (Morrone 2014), exists since the Late Cretaceous. The phylogenetic affinities of Mountain Mesoamerican taxa are ancient South American, and they are distributed mainly in mountain and cloud forests, especially on the slopes receiving the humid oceanic winds between 1200 and 2000 m (Halffter et al. 2019), penetrating occasionally in pine-oak forests. Halffter (1978) noted that the insects following this pattern are relatively scarce, compared to other patterns of the Mexican Transition Zone, but it corresponds to several reptiles and amphibians.

The Mountain Mesoamerican cenocron is equivalent to the Middle American element described by Savage (1982) for the herpetofauna. The author characterized these taxa as primarily tropical, with their closest relatives in Central and South America. He considered that these taxa or their ancestors had more extensive ranges in the Paleogene, when humid warm climates existed as far north as the United States.

Some plant taxa corresponding to the Mountain Mesoamerican cenocron are *Ceratozamia* (Zamiaceae) (González and Vovides 2002) and *Moussonia deppeana* (Gesneriaceae) (Ornelas and González 2014). The latter is a shrub with disjunct distribution in the Mexican Transition Zone (Fig. 5.5), which shows a deep evolutionary split in the Pliocene and has experienced isolation and divergence during interglacial periods (Ornelas and González 2014).

Among arachnids the genus *Bonnetina* (Araneae: Theraphosidae) is a representative of a Mountain Mesoamerican taxon (Ortiz and Francke 2017; Ortiz et al. 2017). A total of 17 species endemic to Mexico has been recognized, and the phylogenetic analysis indicates that the tribe Hapalopini, to which *Bonnetina* belongs, is monophyletic, although the relationships of this genus with the other North American genera are not clearly resolved. The divergence of the three tribes of Theraphosinae might have occurred close to the Cretaceous-Paleocene limit (68–63 m.y.a.), and North American Hapalopini seem to have diverged in the Early Oligocene (30 m.y.a.). Within *Bonnetina*, the *aviae* clade split from the other

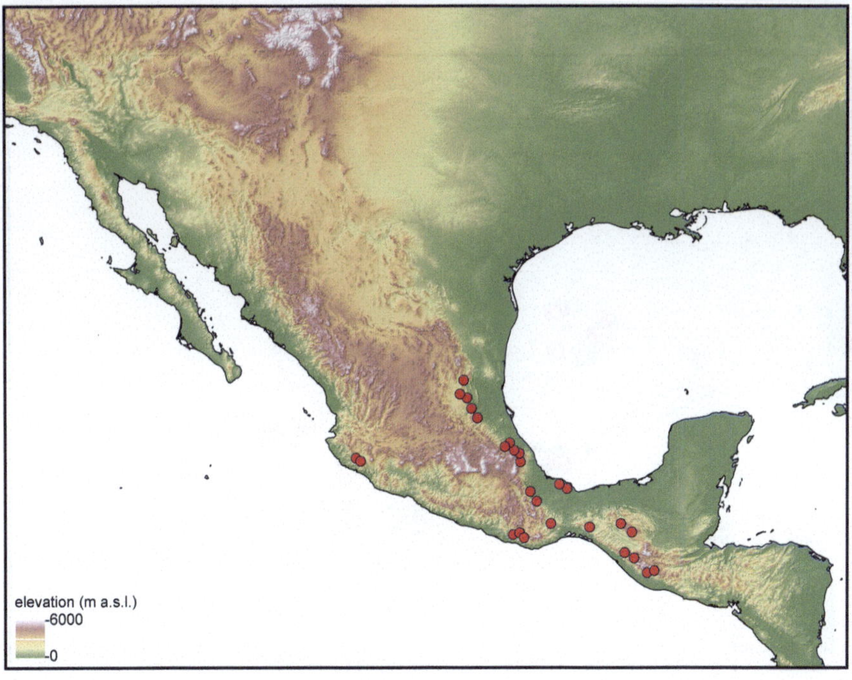

Fig. 5.5 Mexican Plateau distributional pattern: *Moussonia deppeana* (Gesneriaceae), analyzed by Ornelas and González (2014)

species 17 m.y.a., and the *cyaneifemur* and *papalutlensis* clades split 14 m.y.a. Ortiz et al. (2017) postulated that the colonization of Central America by Hapalopini might have occurred during the Oligocene (30 m.y.a.), long before the rise of the Isthmus of Panama.

An insect taxon following the Mountain Mesoamerican pattern is the tribe Proculini (Coleoptera: Passalidae), analyzed by Reyes-Castillo and Halffter (1978). This tribe is endemic to the Neotropics, and in Mexico, it has a predominantly montane distribution, with its highest diversity between 1000 and 2500 m altitude (Fig. 5.6). Based on their phylogenetic affinities with some South American taxa, the authors postulated that these taxa evolved and diversified in the Central American Nucleus, from an ancestor that dispersed there from South America in the Oligocene-Miocene. Several genera of Passalidae are diversified exclusively in the mountains of the Mexican Transition Zone (e.g., *Chondrocephalus*, *Vindex*, *Ogyges*, *Petrejoides*, and *Pseudoarrox*), with some expansion to the lowlands. Some genera show a north-east vicariance (Central American Nucleus and mountains north of the Isthmus of Tehuantepec), whereas other genera show an east-west vicariance (both in the Central American Nucleus and the northern areas). Other beetle taxa that belong to the Mountain Mesoamerican cenocron are the *Platynus degallieri* species group (Coleoptera: Carabidae) (Liebherr 1994) and the genera *Dendroctonus*, *Pantomorus*, and *Plumolepilius* (Coleoptera: Curculionidae) (Salinas-Moreno et al. 2004; Rosas et al. 2011a, b; Barrios-Izás et al. 2016).

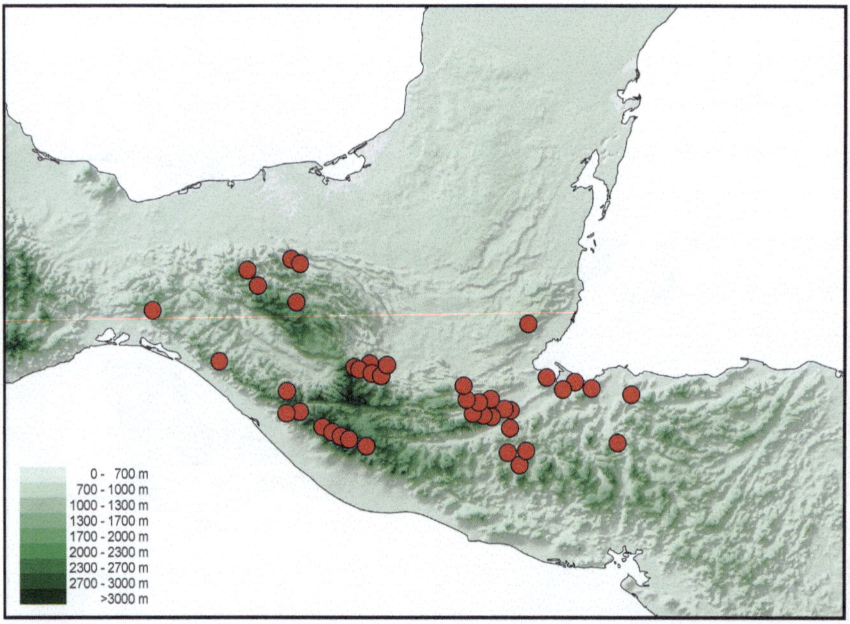

Fig. 5.6 Mountain Mesoamerican distributional pattern: *Proculus* (Coleoptera: Passalidae) (modified from Halffter and Morrone 2017)

An interesting example of a Mountain Mesoamerican beetle taxon is *Ogyges* (Coleoptera: Passalidae), analyzed by Cano et al. (2018). This genus is distributed in Nuclear Central America from Chiapas to northwestern Nicaragua. The phylogenetic analysis indicates that it comprises three species groups, which show allopatric distribution (Fig. 5.7). The *O. championi* species group, with ten species, is distributed in the mountainous system north of the Motozintla-Comaltitlán fault in Chiapas and north of the dry valleys of the Cuilco and Motagua rivers in Guatemala. The *O. laevissimus* species group, including seven species, ranges mostly along the Pacific Volcanic Chain from Guatemala to El Salvador and from southeastern Honduras to northwestern Nicaragua. The *O. crassulus* species group, with ten species, is distributed from northeastern Guatemala to northern Honduras. The Isthmus of Tehuantepec, the Motagua-Cuilco and Motozintla-Comaltitlán sutures zones, the lowland valleys of Colón and Comalí rivers between Nicaragua and Honduras, the Guayape fault system in Honduras, and the dry valleys of Ulúa-Chamelecón-Olancho in Honduras are hypothesized as the vicariance events that affected the geographical distribution of *Ogyges* (Cano et al. 2018).

Some Mountain Mesoamerican bird taxa include *Cyanolyca* (Corvidae) (Bonaccorso 2009), *Buarremon* (Emberizidae) (Navarro-Sigüenza et al. 2008),

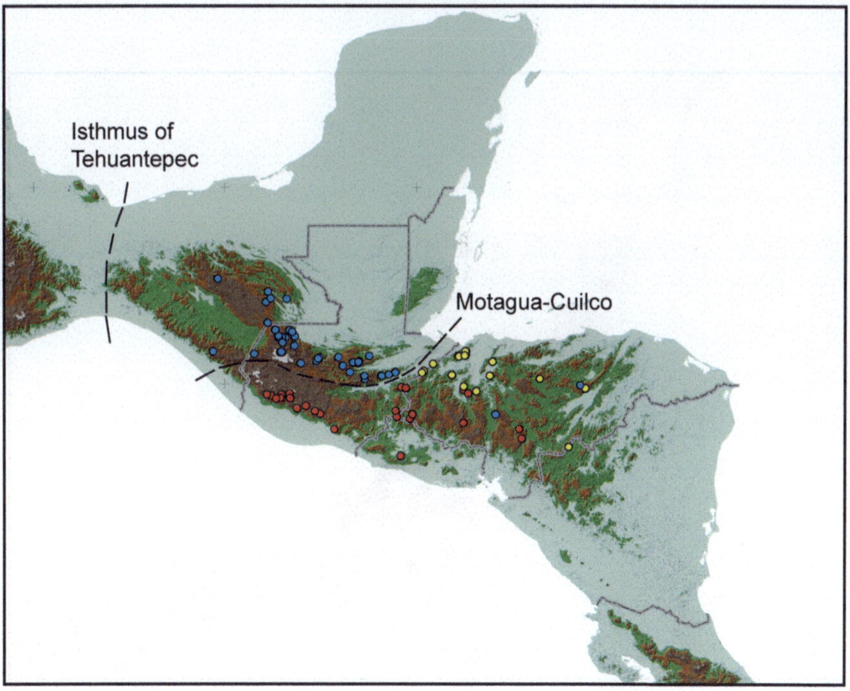

Fig. 5.7 Mountain Mesoamerican distributional pattern: *Ogyges* (Coleoptera: Passalidae) (modified from Cano et al. 2018). Blue circles: *O. championi* species group; red circles: *O. laevissimus* species group; yellow circles: *O. crassulus* species group

Campylorhynchus rufinucha (Troglodytidae) (Vázquez-Miranda et al. 2009), and *Amazilia violiceps* and *A. viridifrons* (Trochilidae) (Rodríguez-Gómez and Ornelas 2015).

5.5 The Nearctic Cenocron: Northern Immigrants

The Nearctic cenocron corresponds to North American taxa that dispersed during the Miocene-Pleistocene to the Sierra Madre Occidental, Sierra Madre Oriental, Sierra Madre del Sur, Transmexican Volcanic Belt, and Chiapas Highlands (Halffter 1964, 1987, 2017; Halffter and Morrone 2017). Some species may have some scattered localities in the Mexican Plateau.

Several plant taxa endemic to the temperate and mountain humid forests of the Mexican Transition Zone (Espinosa Organista et al. 2008) are characteristic of the Nearctic cenocron. Some examples include *Abies* (Pinaceae) and *Pinus* (Pinaceae; Fig. 5.8), which based on fossil evidence seems to have been already present in southern Mexico by the Early Eocene (Perry et al. 2000). Another example is *Nolina parviflora* (Asparagaceae), where divergence time estimates suggest that

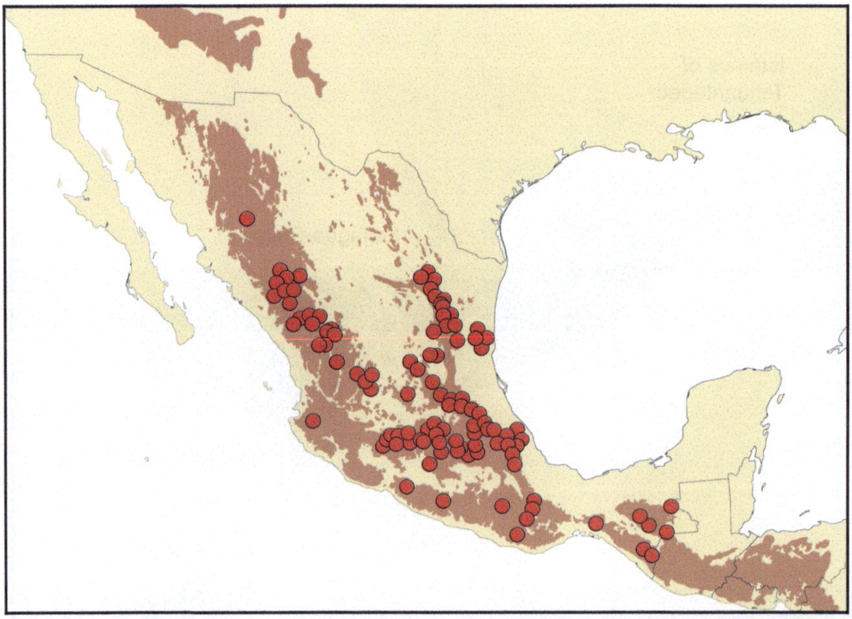

Fig. 5.8 Nearctic distributional pattern: *Pinus teocote* (Pinaceae) (modified from Espinosa Organista et al. 2008)

there is a correlation with periods of uplift of the Transmexican Volcanic Belt that provided barriers that limited gene flow (Ruiz-Sánchez and Specht 2013).

Among arachnids, the genus *Mastigoproctus* (Thelyphonida: Thelyphonidae) (Barrales-Alcalá et al. 2018) is a representative of a Nearctic taxon. Insect taxa assigned to the Nearctic cenocron include *Taeniopoda-Romalea* (Orthoptera: Romaleidae) (De Jesús-Bonilla et al. 2017), *Thanatophilus* (Coleoptera: Silphidae) (Fig. 5.9; Halffter and Morrone 2017), *Elliptoleus* (Coleoptera: Carabidae; Fig. 5.10) (Liebherr 1991, 1994), and *Phalacropsylla* (Siphonaptera: Ctenophthalmidae) (Acosta and Hastriter 2017).

Nearctic vertebrates include *Pseudoeurycea belli* and *Bolitoglossa macrinii* (Amphibia: Plethodontidae; Parra-Olea et al. 2002, 2005, 2010), *Barisia* (Squamata: Anguidae; Bryson and Riddle 2012), *Crotalus intermedius* (Squamata: Viperidae; Bryson et al. 2011), *Plestiodon* (Squamata: Scincidae; Feria-Ortiz et al. 2011), *Melanerpes carolinus* species group (Aves: Picidae; García-Trejo et al. 2009), *Glaucomys volans* (Mammalia: Sciuridae; Kerhoulas and Arbogast 2010), and *Habromys*, *Neotoma*, *Peromyscus*, and *Reithrodontomys* (Mammalia: Cricetidae; Arellano et al. 2003; León-Paniagua et al. 2007; Ávila-Valle et al. 2012; Ordóñez-Garza et al. 2014).

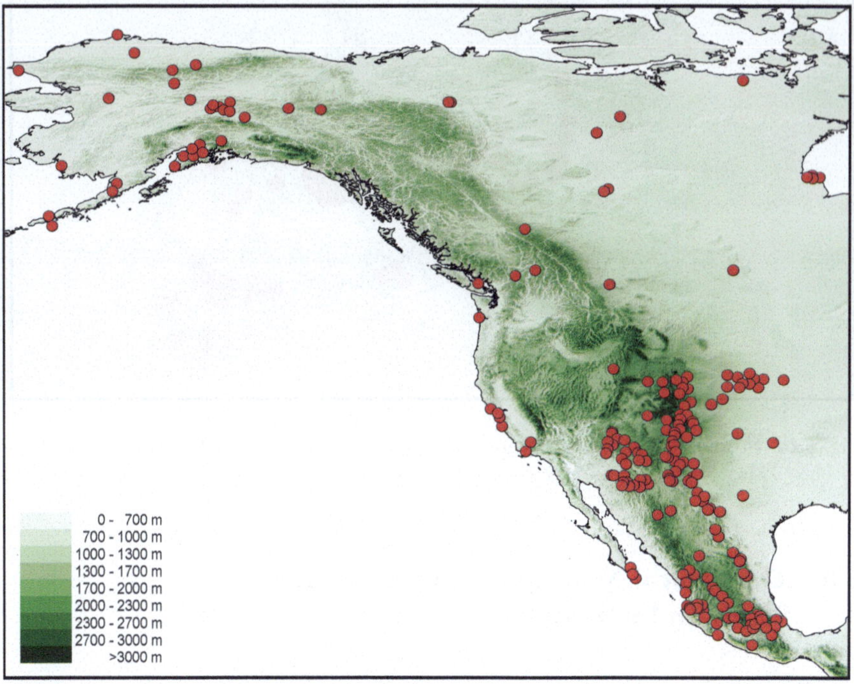

Fig. 5.9 Nearctic distributional pattern: *Thanatophilus* (Coleoptera: Silphidae) (modified from Halffter and Morrone 2017)

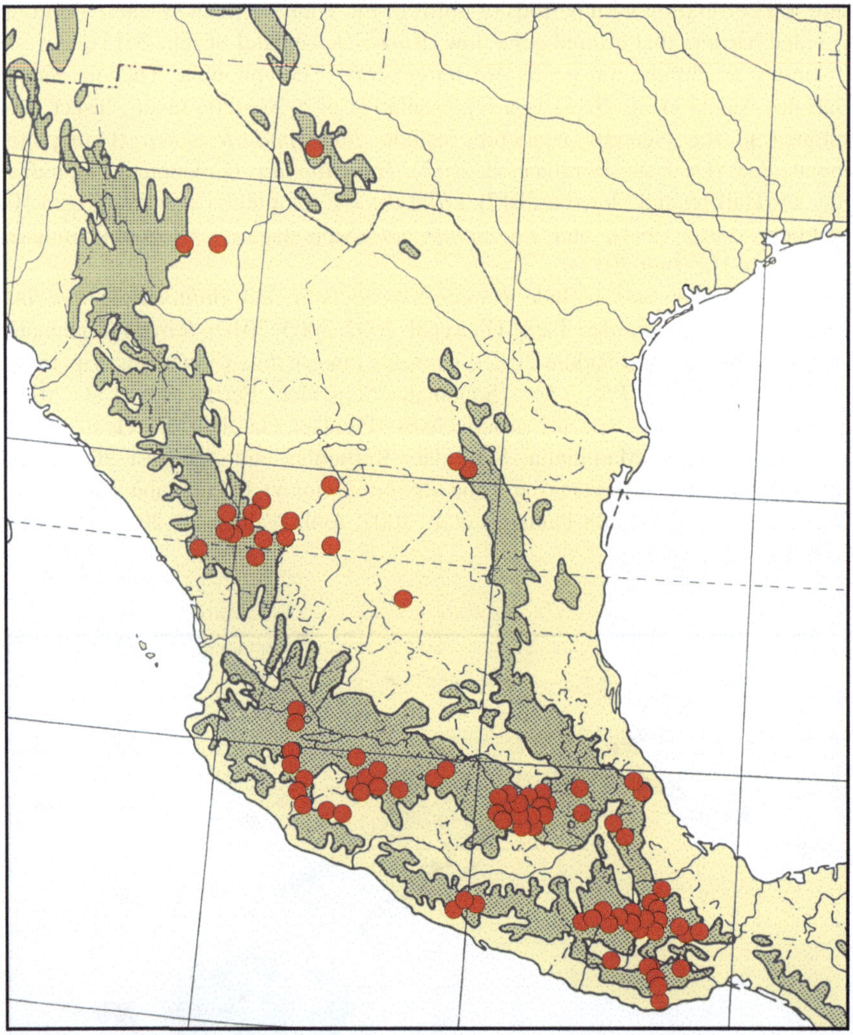

Fig. 5.10 Nearctic distributional pattern: *Elliptoleus* (Coleoptera: Carabidae) (modified from Liebherr 1991). Stippled areas represent temperate forests

5.6 The Typical Neotropical Cenocron: Last Southern Immigrants

The Typical Neotropical cenocron corresponds to taxa that evolved originally in South America, in the Neotropical region. They are represented basically by genera widespread in South America; some Mexican species that have been assigned to these genera may extend to Panama and even reach northern South America (Halffter 1962, 1987). Typical Neotropical species are delimited more ecologically than geographically, because they are found in lowland tropical forests and do not extend to the Mexican Plateau because of its drier conditions (Halffter and Morrone 2017). Along the coasts of the Gulf of Mexico and the Pacific Ocean, Typical Neotropical species extend their distribution to the north, even reaching the southern United States.

The Typical Neotropical cenocron is equivalent to the South American element characterized by Savage (1982) for the herpetofauna. Authors working with mammal taxa have analyzed the Great American Biotic Interchange (GABI) that corresponds to the assemblage of North American and South American biotas that took place when the Panamanian Isthmus was permanently established by the end of the Neogene (Webb 1985; Cody et al. 2010). The precise timing of this event is controversial; some authors restrict the Great American Biotic Interchange to the Pleistocene (e.g., Woodburne 2010), whereas others prefer to include also the Late Miocene to Holocene dispersals (Cione et al. 2015).

Typical Neotropical taxa include *Enckea* (Piperaceae) (Quijano-Abril et al. 2014), *Physalis* (Solanaceae) (Zamora-Tavares et al. 2016), *Heterolinus* and *Homalolinus* (Coleoptera: Staphylinidae) (Márquez and Morrone 2003), *Pselliopus* (Heteroptera: Reduviidae) (Mariño-Pérez et al. 2007), *Anolis sericeus* species complex (Squamata: Dactyloidae) (Gray et al. 2019), *Ergaticus* (Aves: Parulidae) (Barrera-Guzmán et al. 2012), *Chlorospingus ophthalmicus* (Aves: Emberizidae) (Maldonado-Sánchez et al. 2016), and *Handleyomys* (Mammalia: Cricetidae) (Almendra et al. 2018).

Several species of Scarabaeidae (Coleoptera) are characteristic of this pattern, for example, *Canthon cyanellus* (Nolasco-Soto et al. 2017) and *C. indigaceus*, with three subspecies (Fig. 5.11). The Scarabaeid genera *Notiophanaeus*, *Sulcophanaeus*, and *Coprophanaeus* and some species groups of *Phanaeus (Phanaeus)* are other examples of Typical Neotropical Scarabaeidae (Halffter and Morrone 2017). These taxa did not pass the Transmexican Volcanic Belt nor colonize the Mexican Plateau, exhibiting marked vicariance between the coasts of the Pacific and Gulf of Mexico, the Balsas Basin, and the Tehuacán-Cuicatlán Valley.

The Typical Neotropical pattern exhibits different degrees of "penetration" in the Mexican Transition Zone (Halffter and Morrone 2017). In one extreme, there are genera with many species in the transition zone, some of which may have even colonized the Mexican Plateau, as the *Canthon (Glaphyrocanthon) viridis* species group (Fig. 5.12a). Differences with the previous species are notorious, being an exclusively Mexican group, where the genus has speciated by vicariance. In addition to 15 Mexican species, this group has one species widely distributed in North America, which reaches Canada following the eastern coast, and two species in Central America. There are genera with several species in South America that have a single species in Mexico that does not extend north of the Isthmus of Tehuantepec or Chiapas. They correspond to the Neotropical pattern of minimal penetration. One example is *Deltochilum acropyge* (Fig. 5.12b), which is limited

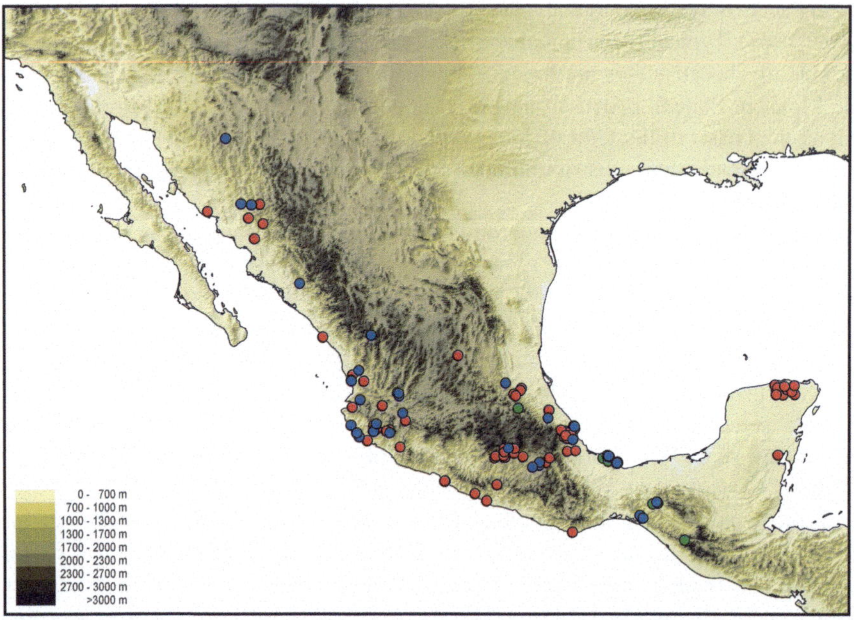

Fig. 5.11 Typical Neotropical distributional pattern with wide penetration: *Canthon indigaceus* (Coleoptera: Scarabaeidae), with three subspecies (modified from Halffter and Morrone 2017). Red circle, *C. indigaceus chevrolati*; green circle: *C. indigaceus chiapas*; blue circle, *C. indigaceus indigaceus*

in Mexico to Laguna Bélgica, Ocozocoautla (Chiapas), and Calakmul (Campeche). Another example of minimal penetration is *Canthon angustatus* (Fig. 5.13a), which is distributed along the northern Andes in South America, and its penetration to Mexico is limited to Bocas de Chajul, in the Lacandonian Forest. There are also taxa with an intermediate penetration in the Mexican Transition Zone. *Canthon (Glaphyrocanthon) subhyalinus* (Fig. 5.13b) is a species found in rainforests of southeastern Mexico, where it is associated closely to monkeys, and apparently correspond to the last expansion of the rainforests in the last 10,000 years (Halffter and Morrone 2017).

The seasonally dry tropical forest represents a vegetation characteristic of the Typical Neotropical cenocron. In order to analyze its origin and expansion in Mexico, Becerra (2005) considered the genus *Bursera* (Burseraceae) to be an appropriate representative. She obtained a time-calibrated phylogenetic hypothesis and reconstructed its biogeographic history, concluding that between 30 and 20 m.y.a. *Bursera* had a rapid radiation. The oldest lineages seem to be those from the Pacific forests, whereas the most recent diverged in southeastern Mexico. De-Nova et al. (2012) obtained a complete phylogenetic hypothesis, concluding that most of the previous estimates have overestimated the divergence dates. They found that although *Bursera* is ca. 50 m.y. old, most of its species are considerably younger,

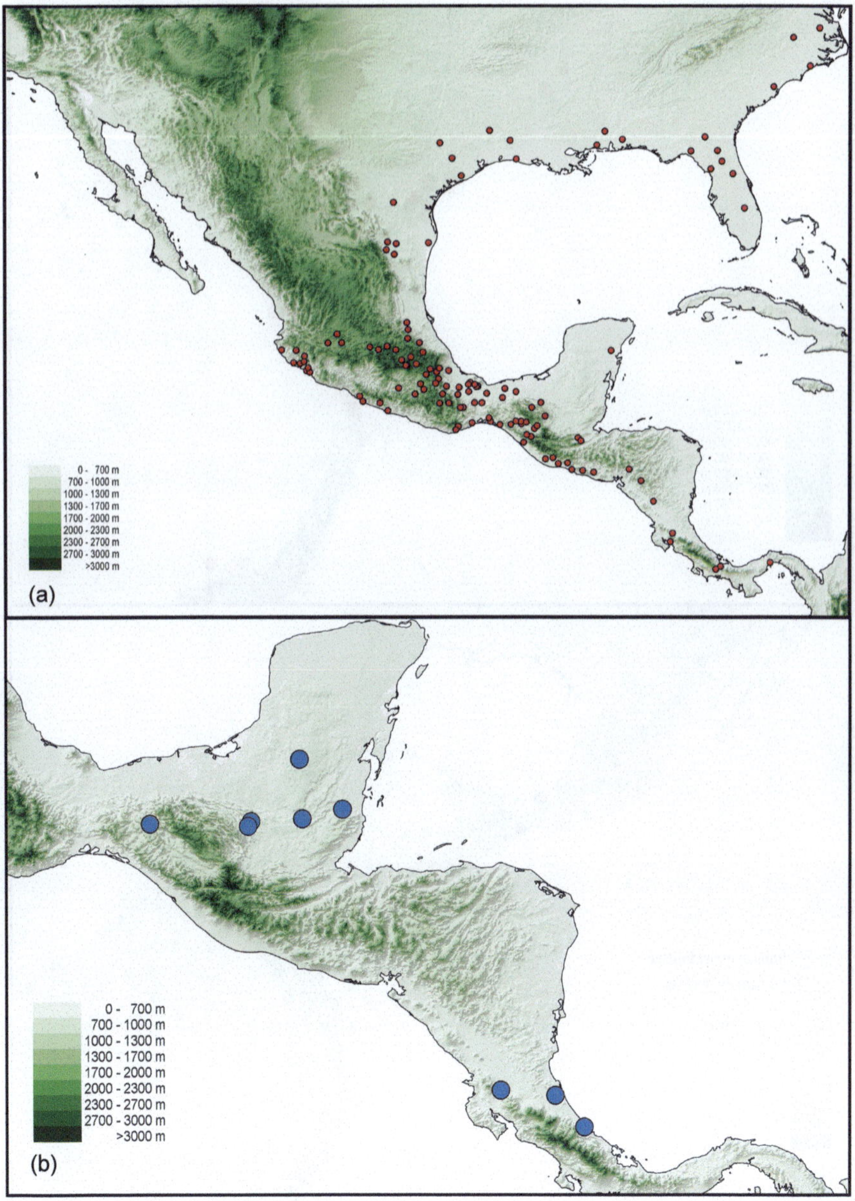

Fig. 5.12 Typical Neotropical distributional pattern. (**a**) Wide penetration, *Canthon (Glaphyrocanthon) viridis* species group (Coleoptera: Scarabaeidae); (**b**) minimal penetration, *Deltochilum acropyge* (Coleoptera: Scarabaeidae) (modified from Halffter and Morrone 2017)

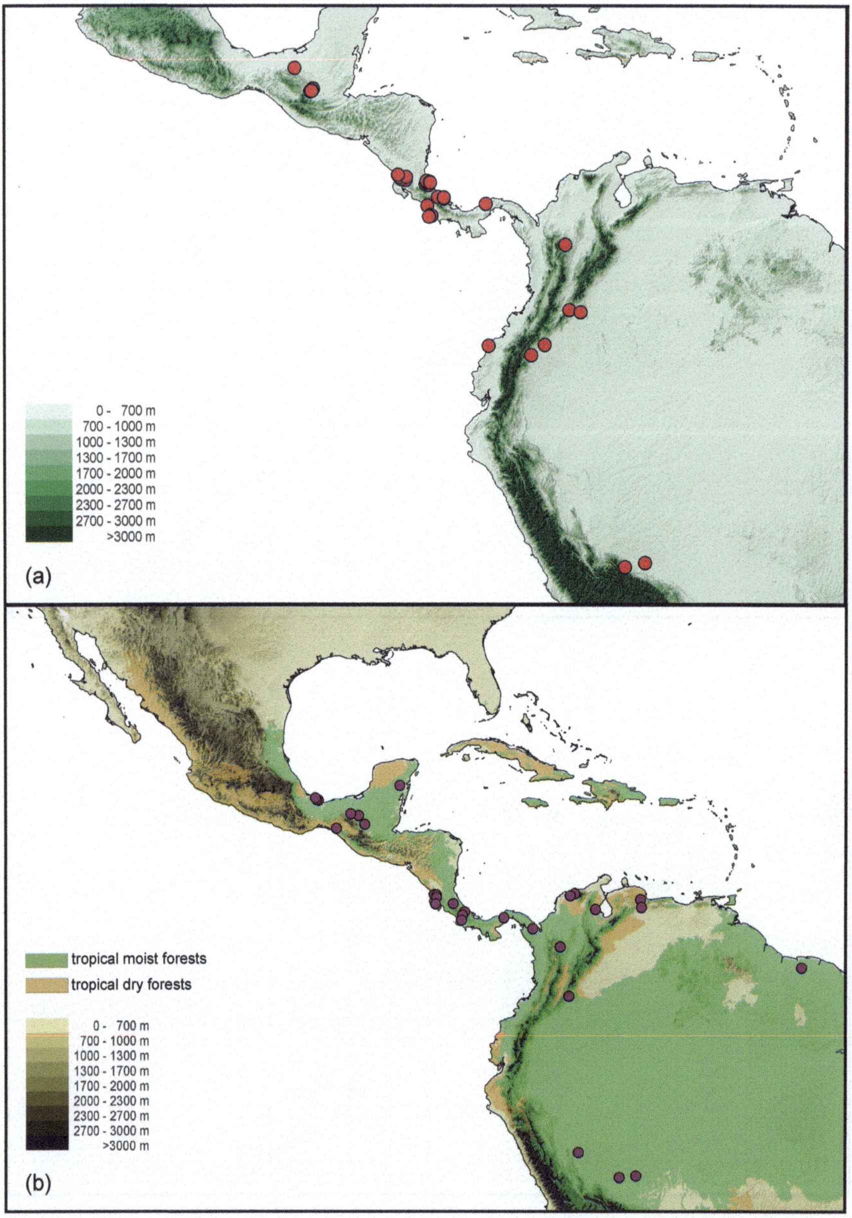

Fig. 5.13 Typical Neotropical distributional pattern. (**a**) Minimal penetration, *Canthon angustatus* (Coleoptera: Scarabaeidae); (**b**) medium penetration, *Canthon (Glaphyrocanthon) subhyalinus* (Coleoptera: Scarabaeidae) (modified from Halffter and Morrone 2017)

and the extant species seemed to have originated between 20 and 17.5 m.y.a., when the Miocene aridification trend reached its peak. This is congruent with the postulated expansion of the seasonally dry tropical forest to northern and central Mexico during the Miocene.

5.7 Assembly of the Cenocrons in the Mexican Transition Zone: The Geobiotic Scenario

The distributional patterns recognized by Halffter and collaborators have been considered to represent cenocrons (Morrone 2005, 2010, 2015a; Zunino 2007; Halffter and Morrone 2017). If the concept of cenocron is applied strictly to a set of taxa that dispersed and integrated into a biota, the Mexican Plateau, Mountain Mesoamerican, Nearctic, and Typical Neotropical qualify clearly as cenocrons, whereas the Paleoamerican pattern represents the original biota of North America. A summary of the biotic development of the Mexican Transition Zone consists of five main stages (Morrone 2015a) and the assembly of three successive horobiotas (Fig. 5.14).

In the Jurassic-Cretaceous, the Paleoamerican biota extended in the southern part of North America that corresponds to what is currently northern Mexico. This biota represents the original (Holarctic) biota of the Mexican Transition Zone, and its representatives have evolved in this area for a long period of time, having diversified into different subpatterns. In the Early Cretaceous, when the Paleoamerican biota extended in what is now western Mexico, there was no connection between North and South America (Fig. 5.15a, b). Paleoamerican taxa have been affected by all the vicariance and biogeographic convergence events that shaped the complex biotic history of the Mexican Transition Zone (Halffter and Morrone 2017).

In the Late Cretaceous-Paleocene, North and South America connected by the proto-Antilles, allowing the Mexican Plateau cenocron to disperse from South America/Gondwana to the Mexican Transition Zone. During the Late Cretaceous-Eocene (Fig. 5.15c–e), the proto-Antilles were situated where Lower Central America is situated in the present time. Then, they drifted eastward to originate the Antillean archipelago, providing a connection that allowed the dispersal of Gondwanic taxa to North America. Mexican Plateau taxa are widely distributed in the Mexican Plateau. In the Paleocene, this connection was interrupted and not re-established until the Miocene (Montes et al. 2015). During this time, marine transgressions covered a large part of South America, from Patagonia to Bolivia and Peru, separating clearly northern-northwestern (Neotropical) from southern-southeastern (Andean) South America (Donato et al. 2003). This isolation has been the main obstacle for the dispersal of Austral taxa, known as Paleantarctic (Jeannel 1942) or Andean (Morrone 2015b), to the Mexican Transition Zone.

The assemblage of Paleoamerican and Mexican Plateau taxa constitutes the Paleogene horobiota of the Mexican Transition Zone, which corresponds basically to the Eocene time-slice. Based on the presence of Neotropical elements, the

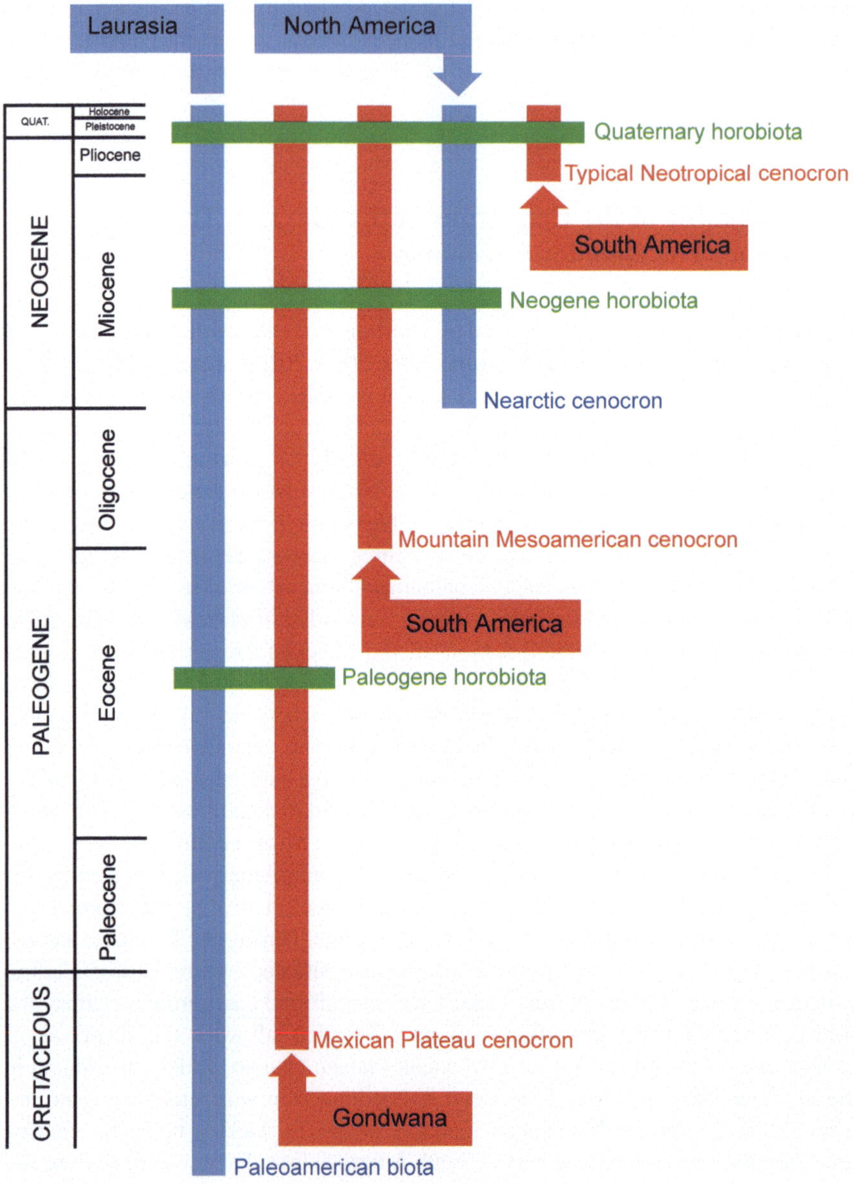

Fig. 5.14 Diagrammatic representation of the biotic assembly of the Mexican Transition Zone, with the successive incorporation of the cenocrons to the original biota and the resulting horobiotas

Mexican Plateau, known as the Chihuahuan Desert biogeographic province, may be considered part of the Mexican Transition Zone; however, I prefer to assign this province to the Nearctic region considering that its biota shows strong biotic relationships to the Sonoran and Baja California provinces.

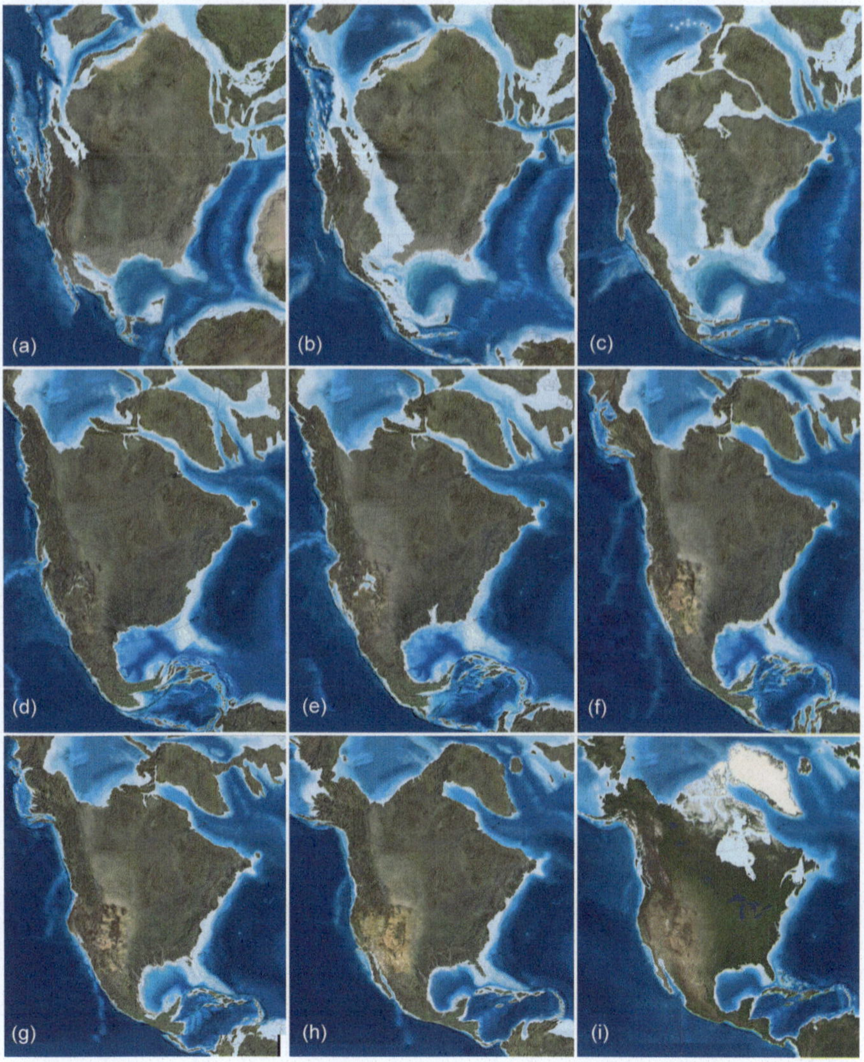

Fig. 5.15 Temporal sequence of paleographical maps of North America. (**a**) Early Cretaceous (130 m.y.a.); (**b**) Early Cretaceous (110 m.y.a.); (**c**) Late Cretaceous (87 m.y.a.); (**d**) Paleocene (60 m.y.a.); (**e**) Eocene (50 m.y.a.); (**f**) Oligocene (35 m.y.a.); (**g**) Miocene (20 m.y.a.); (**h**) Pliocene (5 m.y.a.); (**i**) recent (reproduced with permission of Ronald Blakey, Colorado Plateau Geosystems, Inc.)

In the Oligocene-Miocene (Fig. 5.15f, g), the Mountain Mesoamerican cenocron evolved in the Central American Nucleus. In the Pliocene (Fig. 5.15h), it dispersed northward to the northern part of the Mexican Transition Zone. Mountain Mesoamerican taxa, with ancient South American affinities, evolved in the highlands of Chiapas and Central America north of Lake Nicaragua and then dispersed northward. During the Oligocene began the development of the Transmexican Volcanic Belt, a volcanic area of 1000 km length that runs from the Pacific to the Atlantic Ocean between 18° 30′ N and 21° 30′ N parallels (Ferrari et al. 1999, 2012). Simultaneous to the development of the Transmexican Volcanic Belt from the Miocene to the present, the Mexican Plateau elevated, becoming isolated from southern Mexico by the Sierra Madre Occidental to the west, the Sierra Madre Oriental to the east, and the Transmexican Volcanic Belt to the south. The Mexican Plateau and the Transmexican Volcanic Belt are key geomorphological features that help explain the dispersal/vicariance of the different cenocrons (Corral-Rosas and Morrone 2017). The Mexican Plateau preserves ancient Neotropical lineages, isolated by the mountains of the Mexican Transition Zone. The Sierra Madre Occidental and the Sierra Madre Oriental, united in their southern extremes by the Transmexican Volcanic Belt, and continuing to the south by the Sierra Madre del Sur, are the main area where septentrional taxa dispersed to the south (horizontal colonization; Lobo and Halffter 2000). The Transmexican Volcanic Belt is the most important area for allopatric speciation, due to the repeated episodes of horizontal colonization and vicariance. Halffter and Morrone (2017) considered that in the absence of the Transmexican Volcanic Belt, there would be no Mexican Transition Zone. Its east-west disposition favored allopatric speciation by vicariance, whereas the Sierra Madre Occidental and Sierra Madre Oriental represented dispersal routes. For tropical taxa, the Transmexican Volcanic Belt and the Mexican Plateau represent a clear geographic-climatic barrier. The geological evolution of the Transmexican Volcanic Belt is divided into four episodes, which started in the Miocene. The final episode (Late Pliocene-Holocene) is still active and is characterized by the development of large (> 3500 m) volcanoes in the last 1.5 m.y.a. Geological information on the Transmexican Volcanic Belt was compiled by Ferrusquía-Villafranca (2007), Torres Miranda and Luna (2007), Ferrari et al. (2012), and Mastretta-Yanes et al. (2015).

In the Miocene-Pliocene (Fig. 5.15g, h), the Nearctic cenocron dispersed southward from northern North America to the Mexican Transition Zone, including both Holarctic and Nearctic taxa. Species assigned to this cenocron are generally restricted to areas above 2500–3000 m, being abundant in temperate and cloud forests.

The assemblage of Paleoamerican, Mexican Plateau, Mountain Mesoamerican, and Nearctic taxa constitutes the Neogene horobiota of the Mexican Transition Zone, which corresponds basically to the Miocene time-slice. It is rather similar to the present Mexican Transition Zone, although it still lacks the Typical Neotropical taxa.

In the Pliocene-Pleistocene (Fig. 5.15i), when the Panamanian Isthmus was completed, the Typical Neotropical cenocron dispersed from South America, process that continues to the present. Typical Neotropical taxa are basically distributed in tropical forests of the lowlands and have not been able to invade the highlands. These taxa include both old Gondwanan taxa and younger Neotropical groups. With the re-establishment of the Isthmus of Panama in the Miocene, the Neotropical biota began to disperse northward massively from South America. Montes et al. (2015) have questioned the youth of the generally accepted timing of the re-establishment of the Panamanian Bridge, placing it between 13 and 15 million years ago, not as early as the three million year window generally accepted. To postulate whether a given South American taxon migrated northward during the old connection (prior to the interruption of the Panamanian Bridge) or the later connection (after its re-establishment) is very important. If the reconnection is at the earlier date, it would antedate the uplift of the Mexican Plateau, which otherwise is a barrier to northward dispersal. Consequently, the amount of time that the Panamanian land bridge is interrupted is crucial to any hypothesis about the connections between North and South America.

With this last episode, we have finally the assemblage that currently characterizes the Mexican Transition Zone, with Paleoamerican, Mexican Plateau, Mountain Mesoamerican, Nearctic, and Typical Neotropical taxa. This assemblage represents the Quaternary horobiota of the Mexican Transition Zone, corresponding basically to the Pleistocene time-slice.

5.8 Relevance of the Biotic Assembly of the Mexican Transition Zone

The general geobiotic scenario postulated in this chapter allows to integrate the successive dispersal events and the assembly of the different cenocrons. The Mexican Transition Zone has had a continuous connection with North America since the Mesozoic, allowing the dispersal of septentrional lineages without major barriers, so in some cases, it is difficult to distinguish Paleoamerican and Nearctic taxa, at least without the molecular dating of the lineages (Halffter and Morrone 2017). In contrast, the communication with South America has been interrupted making easier to distinguish the Mexican Plateau, Mountain Mesoamerican, and Typical Neotropical cenocrons. After each of these cenocrons assembled in a new horobiota, vicariance events have led to allopatric speciation, producing the species endemic to the Mexican Transition Zone.

In addition to its evolutionary biogeographic relevance, biotic assembly may be relevant for biodiversity conservation (Halffter and Morrone 2017). For example, Moctezuma et al. (2016) analyzed the influence of ecological and evolutionary factors on communities of copronecrophagous beetles (Coleoptera: Scarabaeidae, Silphidae, and Trogidae) in two mountains of the Transmexican Volcanic Belt with contrasting characteristics, both having areas with well-conserved and disturbed vegetation. They found that both conserved and disturbed areas on the xeric mountain were dominated by species assigned to the Mexican Plateau Paleoamerican subpattern, being this dominance greater in the well-conserved vegetation. On the temperate mountain, the Mexican Plateau Paleoamerican subpattern was dominant in the conserved areas, but the Mountain Paleoamerican subpattern was dominant in the disturbed ones. The authors concluded that on the xeric mountain, vertical colonization has been more important, with predominant Mexican Plateau Paleoamerican species, whereas in the temperate mountain, the process of horizontal colonization has been more important, with mountain lineages (Mountain Paleoamerican and Nearctic) that have the capacity to exploit disturbed habitats. Thus, the biogeographic affinities of the species in terms of the cenocrons to which they belong may have great relevance when analyzing the response to anthropic disturbance.

References

Acosta R, Hastriter MW (2017) A review of the flea genus *Phalacropsylla* Rothschild, 1915 (Siphonaptera, Ctenophthalmidae, Neopsyllinae, Phalacropsyllini) with new host and distributional records. ZooKeys 675:27–43

Almendra AL, González-Cózatl FX, Engstrom MD, Rogers DS (2018) Evolutionary relationships and climatic niche evolution in the genus *Handleyomys* (Sigmodontinae: Oryzomyini). Mol Phylogenet Evol 128:12–25

Arellano E, Rogers D, Cervantes F (2003) Genetic differentiation and phylogenetic relationships among tropical harvest mice (*Reithrodontomys*: subgenus *Aporodon*). J Mammal 84:129–143

Arriaga A, Halffter G, Moreno CE (2012) Biogeographical affinities and species richness of copronecrophagous beetles (Scarabaeoidea) in the southeastern Mexican High Plateau. Rev Mex Biodivers 83:519–529

Ávila-Valle Z, Castro-Campillo A, León-Paniagua L, Salgado-Ugalde I, Navarro-Sigüenza A, Hernández-Baños B, Ramírez-Pulido J (2012) Geographic variation and molecular evidence of the blackish deer mouse complex (*Peromyscus furvus*, Rodentia: Muridae). Mamm Biol 77:166–177

Barragán F, Moreno CE, Escobar F, Bueno-Villegas J, Halffter G (2014) The impact of grazing on dung beetle diversity depends on both biogeographical and ecological context. J Biogeogr 41:1991–2002

Barrales-Alcalá D, Francke OF, Prendini L (2018) Systematic revision of the giant vinegaroons of the *Mastigoproctus giganteus* complex (Thelyphonida: Thelyphonidae) of North America. Bull Am Mus Nat Hist 418:1–62

Barrera-Guzmán AO, Milá B, Sánchez-González LA, Navarro-Sigüenza AG (2012) Speciation in an avian complex endemic to the mountains of Middle America (*Ergaticus*, Aves: Parulidae). Mol Phylogenet Evol 62:907–920

Barrios-Izás MA, Anderson RS, Morrone JJ (2016) A taxonomic monograph of the leaf-litter inhabiting weevil genus *Plumolepilius* new genus (Coleoptera: Curculionidae: Molytinae: Conotrachelini) from Mexico, Guatemala, and El Salvador. Zootaxa 4168:61–91

Becerra JX (2005) Timing the origin and expansion of the Mexican tropical dry forest. Proc Natl Acad Sci 102:10919–10923

Bonaccorso E (2009) Historical biogeography and speciation in Neotropical highlands: molecular phylogenetics of the jay genus *Cyanolyca*. Mol Phylogenet Evol 50:618–632

Bryson R, Riddle B (2012) Tracing the origins of widespread highland species: a case of Neogene diversification across the Mexican sierras in an endemic lizard. Biol J Linn Soc 105:382–394

Bryson R, Murphy R, Graham M, Lathrop A, Lazcano D (2011) Ephemeral Pleistocene woodlands connect the dots for highland rattlesnakes of the *Crotalus intermedius* group. J Biogeogr 38:2299–2310

Cano EB, Schuster JC, Morrone JJ (2018) Phylogenetics of *Ogyges* Kaup and the biogeography of Nuclear Central America (Coleoptera: Passalidae). ZooKeys 737:81–111

Catalano SA, Vilardi JC, Tosto D, Saidman BO (2008) Molecular phylogeny and diversification history of *Prosopis* (Fabaceae: Mimosoideae). Biol J Linn Soc 93:621–640

Cione AL, Gasparini GM, Soibelzon E, Soibelzon LH, Tonni EP (2015) The great American biotic interchange: a South American perspective. Springer, Dordrecht

Cody S, Richardson JE, Rulf V, Ellis C, Pennigton RT (2010) The great American biotic interchange revisited. Ecography 33:326–332

Corral-Rosas V, Morrone JJ (2017) Analyzing the assembly of cenocrons in the Mexican transition zone through a time-sliced cladistic biogeographic analysis. Austr Syst Bot 29:489–501

De Jesús-Bonilla VS, Barrientos-Lozano L, Zaldívar-Riverón A (2017) Sequence-based species delineation and molecular phylogenetics of the transitional Nearctic-Neotropical grasshopper genus *Taeniopoda* (Orthoptera, Romaleidae). Syst Biodivers 15:600–617

De-Nova JA, Medina R, Montero JC, Weeks A, Rosell JA, Olson ME, Eguiarte LE, Magallón S (2012) Insights into the historical construction of species-rich Mesoamerican seasonally dry tropical forests: the diversification of *Bursera* (Burseraceae, Sapindales). New Phytol 193:276–287

Diamond JM (1975) Assembly of species communities. In: Cody ML, Diamond JM (eds) Ecology and evolution of communities. Harvard University Press, Cambridge, MA, pp 342–444

Donato M, Posadas P, Miranda Esquivel DR, Ortiz Jaureguizar E, Cladera G (2003) Historical biogeography of the Andean region: evidence from Listroderina (Coleoptera: Curculionidae: Rhytirrhinini) in the context of the South American geobiotic scenario. Biol J Linn Soc 80:339–352

Espinosa Organista D, Ocegueda Cruz S, Aguilar Zúñiga C, Flores Villela O, Llorente-Bousquets J (2008) El conocimiento biogeográfico de las especies y su regionalización natural. In: Sarukhán J (ed.), Capital natural de México. Vol. I. Conocimiento actual de la biodiversidad, Conabio, Mexico City, pp. 33–65

Feria-Ortiz M, Manríquez-Morán N, Nieto-Montes de Oca A (2011) Species limits based on mtDNA and morphological data in the polytypic species *Plestiodon brevirostris* (Squamata: Scincidae). Herpetol Monogr 25:25–51

Ferrari L, López-Martínez M, Aguirre-Díaz G, Carrasco-Nuñez G (1999) Space-time patterns of Cenozoic arc volcanism in Central Mexico: from the Sierra Madre occidental to the Mexican Volcanic Belt. Geology 27:303–306

Ferrari L, Orozco-Esquivel T, Manea V, Manea M (2012) The dynamic history of the trans-Mexican volcanic belt and the Mexico subduction zone. Tectonophysics 522–523:122–149

Ferro I, Morrone JJ (2014) Biogeographic transition zones: a search for conceptual synthesis. Biol J Linn Soc 113:1–12

Ferrusquía-Villafranca I (2007) Ensayo sobre la caracterización y significación biológica. In: Luna I, Morrone JJ, Espinosa D (eds) Biodiversidad de la Faja Volcánica Transmexicana. Universidad Nacional Autónoma de México, Mexico City, pp 7–23

García-Trejo E, Espinosa de los Monteros A, Arizmendi MC, Navarro-Sigüenza A (2009) Molecular systematics of the red-bellied and golden-fronted woodpeckers. Condor 111:442–452

González D, Vovides A (2002) Low intralineage divergence in *Ceratozamia* (Zamiaceae) detected with nuclear ribosomal DNA ITS and chloroplast DNA trnL-F-non-coding region. Syst Bot 27:654–661

Gray LN, Barley AJ, Poe S, Thomson RC, Nieto-Montes de Oca A, Wang IJ (2019) Phylogeography of a widespread lizard complex reflects patterns of both geographic and ecological isolation. Mol Ecol 28:644–657

Gutiérrez-Ortega JS, Salinas-Rodríguez MM, Martínez JF, Molina-Freaner F, Pérez-Farrera MA, Vovides AP, Matsuki Y, Suyama Y, Ohsawa TA, Watano Y, Kajita T (2018) The phylogeography of the cycad genus *Dioon* (Zamiaceae) clarifies its Cenozoic expansion and diversification in the Mexican transition zone. Ann Bot 121:535–548

Halffter G (1962) Explicación preliminar de la distribución geográfica de los Scarabaeidae mexicanos. Acta Zool Mex 5:1–17

Halffter G (1964) La entomofauna americana, ideas acerca de su origen y distribución. Folia Entomol Mex 6:1–108

Halffter G (1976) Distribución de los insectos en la Zona de Transición Mexicana: Relaciones con la entomofauna de Norteamérica. Folia Entomol Mex 35:1–64

Halffter G (1978) Un nuevo patrón de dispersión en la Zona de Transición Mexicana: El mesoamericano de montaña. Folia Entomol Mex 39–40:219–222

Halffter G (1987) Biogeography of the montane entomofauna of Mexico and Central America. Annu Rev Entomol 32:95–114
Halffter G (2017) La Zona de Transición Mexicana y la megadiversidad de México: Del marco histórico a la riqueza actual. Dugesiana 24:78–89
Halffter G, Morrone JJ (2017) An analytical review of Halffter's Mexican transition zone, and its relevance for evolutionary biogeography, ecology and biogeographical regionalization. Zootaxa 4226:1–56
Halffter G, Verdú JR, Márquez J, Moreno CE (2008) Biogeographical analysis of Scarabaeinae and Geotrupinae along a transect in Central Mexico (Coleoptera, Scarabaeoidea). Fragm Entomol 40:273–322
Halffter G, Zunino M, Moctezuma V, Sánchez-Huerta JL (2019) The integration of the distributional patterns in the Mexican Transition Zone: phyletic, paleogeographic and ecological factors of a case study. Zootaxa 4586:1–34
Hrbek T, Seckinger J, Meyer A (2007) A phylogenetic and biogeographic perspective on the evolution of poeciliid fishes. Mol Phylogenet Evol 43:986–998
Jeannel R (1942) La génese des faunes terrestres: Eléments de biogéographie. Presses Universitaires de France, Paris
Kerhoulas N, Arbogast B (2010) Molecular systematics and Pleistocene biogeography of Mesoamerican flying squirrels. J Mammal 91:654–667
Lavin M, Luckow M (1993) Origins and relationships of tropical North America in the context of the boreotropical hypothesis. Amer J Bot 80:1–14
León-Paniagua L, Navarro-Sigüenza A, Hernández-Baños B, Morales J (2007) Diversification of the arboreal mice of the genus *Habromys* (Rodentia: Cricetidae: Neotominae) in the Mesoamerican highlands. Mol Phylogenet Evol 42:653–664
Liebherr JK (1991) Phylogeny and revision of the *Anchomenus* clade: the genera *Tetraleucus, Anchomenus, Sericoda* and *Elliptoleus* (Coleoptera: Carabidae: Platynini). Bull Am Mus Nat Hist 202:1–163
Liebherr JK (1994) Biogeographic patterns of montane Mexican and Central American Carabidae (Coleoptera). Can Entomol 126:841–860
Lobo JM, Halffter G (2000) Biogeographical and ecological factors affecting the altitudinal variation of mountainous communities of coprophagous beetles (Coleoptera: Scarabaeoidea): a comparative study. Ann Entomol Soc Am 93:115–126
Maldonado-Sánchez D, Gutiérrez-Rodríguez C, Ornelas JF (2016) Genetic divergence in the common bush-tanager *Chlorospingus ophthalmicus* (Aves: Emberizidae) throughout Mexican cloud forests: the role of geography, ecology and Pleistocene climatic fluctuations. Mol Phylogenet Evol 99:76–88
Mariño-Pérez R, Brailovsky H, Morrone JJ (2007) Análisis panbiogeográfico de las especies mexicanas de *Pselliopus* Bergroth (Hemiptera: Heteroptera: Reduviidae: Harpactorinae). Acta Zool Mex 23:77–88
Márquez J, Morrone JJ (2003) Análisis panbiogeográfico de las especies de *Heterolinus* y *Homalolinus* (Coleoptera: Staphylinidae: Xantholinini). Acta Zool Mex 90:15–25
Mastretta-Yanes A, Moreno-Letelier A, Piñero D, Jorgensen TH, Emerson BC (2015) Biodiversity in the Mexican highlands and the interaction of geology, geography and climate within the trans-Mexican volcanic belt. J Biogeogr 42:1586–1600
Moctezuma V, Halffter G (2019) New biogeographical makeup for colonisation of the Baja California Peninsula, with the description of a new *Onthophagus* (Coleoptera: Scarabaeidae: Scarabaeinae). J Nat Hist 53:2057–2071
Moctezuma V, Halffter G, Escobar F (2016) Response of copronecrophagous beetle communities to habitat disturbance in two mountains of the Mexican Transition Zone: influence of historical and ecological factors. J Inst Conserv 20:945–956
Montes C, Cardona A, Jaramillo C, Pardo A, Silva JC, Valencia V, Ayala C, Pérez-Angel LC, Rodríguez-Parra LA, Ramírez JC, Niño H (2015) Middle Miocene closure of the Central American seaway. Science 348:226–228
Morrone JJ (2005) Hacia una síntesis biogeográfica de México. Rev Mex Biodivers 76:207–252

Morrone JJ (2010) Fundamental biogeographic patterns across the Mexican transition zone: an evolutionary approach. Ecography 33:355–361

Morrone JJ (2014) Biogeographical regionalisation of the Neotropical region. Zootaxa 3782:1–110

Morrone JJ (2015a) Halffter's Mexican transition zone (1962–2014), cenocrons and evolutionary biogeography. J Zool Syst Evol Res 53:249–257

Morrone JJ (2015b) Biogeographical regionalisation of the Andean region. Zootaxa 3936:207–236

Navarro-Sigüenza A, Peterson AT, Nyari A, García-Derás G, García Moreno J (2008) Phylogeography of the *Buarremon* brush-finch complex (Aves, Emberizidae) in Mesoamerica. Mol Phylogenet Evol 47:21–35

Nolasco-Soto J, González-Astorga J, Espinosa de los Monteros A, Galente-Patiño E, Favila ME (2017) Phylogeographic structure of *Canthon cyanellus* (Coleoptera: Scarabaeidae), a Neotropical dung beetle in the Mexican Transition Zone: insights on its origin and the impacts of Pleistocene fluctuations on population dynamics. Mol Phylogenet Evol 109:180–190

Ordóñez-Garza N, Thompson C, Unkefer M, Edwards C, Owen J, Bradley R (2014) Systematics of the *Neotoma mexicana* species group (Mammalia: Rodentia: Cricetidae) in Mesoamerica: new molecular evidence on the status and relationships of *N. ferruginea*. Proc Biol Soc Wash 127:518–532

Ornelas J, González C (2014) Interglacial genetic diversification of *Moussonia deppeana* (Gesneriaceae), a hummingbird-pollinated, cloud forest shrub in northern Mesoamerica. Mol Ecol 23:4119–4136

Ortiz D, Francke OF (2017) Reconciling morphological and molecular systematics in tarantulas (Araneae: Theraphosidae): revision of the Mexican endemic genus *Bonnetina*. Zool J Linnean Soc 180:819–886

Ortiz D, Francke OF, Bond JE (2017) A tangle of forms and phylogeny: extensive morphological homoplasy and molecular clock heterogeneity in *Bonnetina* and related tarantulas. Mol Phylogenet Evol 127:55–73

Parenti LR (1981) A phylogenetic and biogeographic analysis of Cyprinodontiform fishes (Teleostei, Atherinomorpha). Bull Am Mus Nat Hist 168:335–557

Parra-Olea G, García-París M, Wake D (2002) Phylogenetic relationships among the salamanders of the *Bolitoglossa macrinii* species group (Amphibia: Plethodontidae), with descriptions of two new species from Oaxaca (Mexico). J Herpetol 36:356–366

Parra-Olea G, García-París M, Papebfuss TJ, Wake D (2005) Systematics of the *Pseudoeurycea bellii* (Caudata: Plethodontidae) species complex. Herpetologica 61:145–158

Parra-Olea G, Rovito S, Márquez-Valdelamar L, Cruz G, Murrieta-Galindo R, Wake D (2010) A new species of *Pseudoeurycea* from the cloud forest in Veracruz, Mexico. Zootaxa 2725:57–68

Perry JP, Graham A, Richardson DM (2000) The history of pines in Mexico and Central America. In: Richardson DM (ed) Ecology and biogeography of *Pinus*. Cambridge University Press, Cambridge, pp 137–149

Pulido-Herrera LA, Zunino M (2007) Catálogo preliminar de los Onthophagini de América (Coleoptera: Scarabaeinae). In: Zunino M, Melic A (eds) Escarabajos, diversidad y conservación biológica: Ensayos en homenaje a Gonzalo Halffter,. Monografías 3er. Milenio M3M. Sociedad Entomológica Aragonesa, Zaragoza, pp 93–129

Quijano-Abril MA, Mejía-Franco FG, Callejas-Posadas R (2014) Análisis panbiogeográfico de *Enckea* (Piperaceae), un pequeño clado de bosques secos en la filogenia de un gran género de bosques tropicales. Rev Mex Biodivers 86:98–107

Rahbek C, Borregaard MK, Colwell RK, Dalsgaard B, Holt BG, Morueta-Holme N, Nogues-Bravo D, Whittaker RJ, Fjeldsa J (2019) Humboldt's enigma: what causes global patterns of mountain biodiversity? Science 365:1108–1113

Reyes-Castillo P, Halffter G (1978) Análisis de la distribución de la tribu Proculini (Coleoptera, Passalidae). Folia Entomol Mex 39–40:222–226

Reznick DN, Furnese AI, Meredith RW, Springer MS (2017) The origin and biogeographic diversification of fishes in the family Poeciliidae. PLoS One 12:1–20

Rodríguez-Gómez F, Ornelas JF (2015) At the passing gate: past introgression in the process of species formation between *Amazilia violiceps* and *A. viridifrons* hummingbirds along the Mexican Transition Zone. J Biogeogr 42:1305–1318

Rosas MV, Morrone JJ, del Río MG, Lanteri AA (2011a) Phylogenetic analysis of the *Pantomorus-Naupactus* complex (Coleoptera: Curculionidae: Entiminae) from North and Central America. Zootaxa 2780:1–19

Rosas MV, del Río MG, Lanteri AA, Morrone JJ (2011b) Track analysis of the North and Central American species of the *Pantomorus-Naupactus* complex (Coleoptera: Curculionidae). J Zool Syst Evol Res 49:309–314

Rosen DE (1960) Middle-American poeciliid fishes of the genus *Xiphophorus*. Bull Florida St Mus Biol Sci 5:57–242

Rosen DE (1979) Fishes of the uplands and intermontane basins: revisionary studies and comparative geography. Bull Am Mus Nat Hist 162:1–176

Ruiz-Sánchez E, Specht C (2013) Influence of the geological history of the trans-Mexican volcanic belt on the diversification of *Nolina parviflora* (Asparagaceae: Nolinoideae). J Biogeogr 40:1336–1347

Rzedowski J (1988) Análisis de la distribución geográfica del complejo *Prosopis* (Leguminosae, Mimosoideae) en Norteamérica. Acta Bot Mex 3:7–19

Salinas-Moreno Y, Mendoza MG, Barrios MA, Cisneros A, Macías-Sámano J, Zúñiga G (2004) Areography of the genus *Dendroctonus* (Coleoptera: Curculionidae: Scolytinae) in Mexico. J Biogeogr 31:1163–1177

Savage JM (1982) The enigma of the Central American herpetofauna: dispersals or vicariance? Ann Missouri Bot Gard 69:464–547

Sites JW, Archie JW, Cole CJ, Flores Villela O (1992) A review of the phylogenetic hypotheses of the genus *Sceloporus* (Phrynosomatidae): implications for ecological and evolutionary studies. Bull Am Mus Nat Hist 213:1–110

Torres Miranda A, Luna I (2007) Hacia una síntesis panbiogeográfica. In: Luna I, Morrone JJ, Espinosa D (eds) Biodiversidad de la Faja Volcánica Transmexicana. Universidad Nacional Autónoma de México, Mexico City, pp 502–514

Vázquez-Miranda H, Navarro-Sigüenza A, Omland K (2009) Phylogeography of the rufous-naped wren (*Campylorhynchus rufinucha*): speciation and hybridization in Mesoamerica. Auk 126:765–768

Webb SD (1985) Faunal interchange between North and South America. Acta Zool Fenn 170:177–178

Weeks BC, Claramunt S, Cracraft J (2016) Integrating systematics and biogeography to disentangle the roles of history and ecology in biotic assembly. J Biogeogr 43:1546–1559

Whittaker RJ, Fernández-Palacios JM (2007) Island biogeography: ecology, evolution, and conservation, 2nd edn. Oxford University Press, Oxford

Woodburne MO (2010) The great American biotic interchange: dispersals, tectonics, climate, sea level and holding pens. J Mamm Evol 17:245–264

Yetman D (2007) The great cacti: ethnobotany and biogeography. The University of Arizona Press, Tucson, AZ

Zamora-Tavares MP, Martínez M, Magallón S, Guzmán-Dávalos L, Vargas-Ponce O (2016) *Physalis* and physaloids: a recent and complex evolutionary history. Mol Phylogenet Evol 100:41–50

Zunino M (2007) Latinoamérica ante las ciencias de la naturaleza y del medio ambiente: Materiales para una reflexión histórica. Acta Zool Mex 23:181–190

Zunino M, Halffter G (1988) Análisis taxonómico, ecológico y biogeográfico de un grupo americano de *Onthophagus* (Coleoptera: Scarabaeidae). Mus Reg Scienze Natur Monogr 9:1–211

Chapter 6
Perspectives

> *Don't adventures ever have an end? I suppose not. Someone else always has to carry on the story.*
>
> *John R. R. Tolkien (1854), The Lord of the Rings*

Abstract Halffter's theory of biotic assembly of the Mexican Transition Zone is a coherent set of hypotheses corroborated by numerous studies on plant and animal taxa, which may serve as a model to analyze the other major transition zones of the world (Saharo-Arabian, Chinese, South American, and Indo-Malayan), as well as transition zones at smaller scales. Perspectives of the study of biogeographic transition zones in evolutionary and ecological biogeography include the application of the concepts of cenocrons and horobiotas to different systems, e.g., mountains, islands, habitat patches, etc. The development of a truly integrative biogeography remains a challenge for the future.

6.1 Introduction

Halffter's theory on the Mexican Transition Zone was formulated half a century ago, although it was two decades later (Halffter 1987) that it acquired a complete formalization. Halffter's hypotheses have been tested by several authors, and the historical development of the theory has been analyzed (Reyes-Castillo 2003; Morrone 2015a). A critical review (Halffter and Morrone 2017) highlighted the relevance of sound phylogenetic analyses, lineage dating, and an adequate knowledge of the geological history to continue testing and refining the theory. Until recent years, phylogenetic analyses were based exclusively on morphological data, but molecular analyses and the new methods of phylogenetic inference have provided new ways of formulating and testing biogeographic hypotheses, specially for dating the lineages assembled in transition zones.

J. J. Morrone, *The Mexican Transition Zone*, https://doi.org/10.1007/978-3-030-47917-6_6

In the next years, I hope new analyses will be undertaken on the Mexican Transition Zone. Studying new taxa is critical, because the vast majority of the published studies have been based on insects (mainly Coleoptera), vertebrates, and plants. Lineage dating analyses are still scarce; they are critical for evaluating the time of assemblage of the cenocrons in the Mexican Transition Zone and even for discovering new cenocrons. Ecological biogeographic studies represent an area where I would expect important advances.

In addition to the Mexican Transition Zone, there are other four transition zones already recognized in the world (Morrone 2015b): South American, Saharo-Arabian, Chinese, and Indo-Malayan (Wallacea). Transition zones are specially important, because they can be deconstructed into their constituting cenocrons, in order to analyze the biotic assembly of taxa with different dispersal capacities, speciation modes, and ecological inertia (Halffter and Morrone 2017). These major transition zones are very different, for example, the South American and Chinese Transition Zones are mountainous as the Mexican Transition Zone, but the Saharo-Arabian Transition Zone is desertic and Wallacea is scattered over several islands. I hope my review encourages other researchers to analyze the biotic assembly in these transition zones.

6.2 Evolutionary Biogeography

My perspective of evolutionary biogeography (Morrone 2009) incorporates both the processes of dispersal and vicariance, as previously proposed by Reig (1962, 1981), Savage (1966, 1982), and Halffter (1987). In contrast to other authors that during the last decades of the twentieth century have crusaded for either dispersal or vicariance, these authors anticipated the dispersal-vicariance model (Morrone 2003, 2011) and, instead of assuming that dispersal or vicariance is the only driver of biotic evolution, considered that both processes are relevant and should be considered in the analyses (Brooks 2004; Lieberman 2004; Sanmartín and Ronquist 2004; Riddle et al. 2008; Crisci and Katinas 2009). Dispersal occurs normally and is a prerequisite for vicariance, but also after vicariance dispersal occurs and obliterates the existent patterns. Once biotas have been identified as either areas of endemism or generalized tracks, dating the cenocrons allows to analyze biotic assembly (Morrone 2009).

Cenocrons constitute testable hypotheses (Lobo 2007; Morrone 2015a; Halffter and Morrone 2017). Further studies are needed to refine or falsify them, for example, dating selected lineages and examining their phylogenetic placement and the distribution of their related taxa, and even to discover new cenocrons. If hypotheses on cenocrons are available for a given area, it would be possible to undertake a time-sliced cladistic biogeographic analysis (Corral-Rosas and Morrone 2017). For example, in a case where two cenocrons incorporated to the biota distributed in a given area, three different time-slices or horobiotas may be identified. The oldest time-slice would correspond to the taxa belonging to the original biota, the intermediate time-slice to the taxa belonging to the original biota + the taxa belonging to the first cenocron, and the most recent time-slice to all the taxa together. The separate cladistic biogeographic analyses for the different time-slices could help understand the way vicariance has affected these successive horobiotas.

6.3 Ecological Biogeography

Biogeographic transition zones may be also analyzed from the viewpoint of ecological biogeography (e.g., Lobo and Halffter 2000; Ruggiero and Ezcurra 2003; Halffter et al. 2008). For example, studies of altitudinal variation in richness and composition of beetle communities by Lobo and Halffter (2000) and Halffter et al. (2008), among others, have shown that taxa belonging to different cenocrons show different distributional patterns. Mountain biotas show different biotic assemblages at different altitudes, and when these assemblages are deconstructed into their cenocrons, the patterns become clearer than well all species are analyzed together (Ferro et al. 2017). Lobo and Halffter (2000) found in Cofre de Perote (Veracruz, Mexico) that high altitude communities were dominated by Mountain Paleoamerican or Nearctic species, whereas lowland communities were dominated by Typical Neotropical species.

The identification of cenocrons may be applied profitable in macroecology, the study of the division of food and space among species at large spatial and temporal scales (Brown 1995). If taxa belonging to the same cenocron share some adaptations (Lobo 2007; Halffter and Morrone 2017), it could be expected that their recognition helps analyze large-scale multispecies ecological patterns.

We may also use cenocrons to predict the answer of communities to deforestation and climate change. For example, Escobar et al. (2007) hypothesized that the impact of human activities on scarabaeid communities depends on their biogeographic history (e.g., the cenocron to which they belong). In communities where Typical Neotropical taxa are predominant, deforestation may have drastic effects in richness and abundance, but at higher altitudes with predominance of Nearctic lineages, the effect of deforestation may be less critical (Moctezuma et al. 2016). The reason is that taxa that evolved originally in tropical forest conditions (Typical Neotropical cenocron) seem to have fewer possibilities to adapt to the loss of tropical forests. In contrast, Nearctic taxa that evolved in grasslands and other open habitats are not affected much by the loss of trees. The biogeographic history of a taxon and the particular cenocron to which it belongs may help predict the type of response to a future environmental change.

The impact of climate change in future scenarios has been analyzed by several recent studies. For example, Aguado-Bautista and Escalante (2015) found that Mexican terrestrial mammals are susceptible to loss of patterns of endemicity, geographic displacement and area reduction in future scenarios of climatic change. Some of the species that the authors predict that will be affected by global warming seem to belong to the Nearctic cenocron (e.g., rodents and Soricidae); however, an analysis taking into account explicitly the cenocron to which each taxon belongs would help find whether they behave in a similar or a different way with respect to global warming.

6.4 Integrative Biogeography

Integrative biogeography should go beyond the separate compartments of evolutionary and ecological biogeography (Lieberman 2003; Crisci and Katinas 2009; Morrone 2009; Weeks et al. 2016). Developing the bases for such integration is beyond the scope of this book, but I hope my review of the biotic assembly in the Mexican Transition Zone can stimulate others to undertake such endeavor. In order to provide a starting point for discussing the possible integration of evolutionary and ecological biogeography, I suggest that both taxonomic and biotic phenomena occur in evolutionary and ecological spaces. This means that the orthodox distinction between ecological and evolutionary biogeography has to do more with a perspective along a single axis than with two different disciplines (Fig. 6.1), where both perspectives may be applied to particular taxa or to entire biotas. Any biogeographic study develops in one of these quadrants, and then may move to another; for example, from the niche modeling of a species (taxon/ecological) to the history of a lineage (taxon/evolutionary), or from a macroecological analysis (biota/ecological) to reconstructing the biotic history of an area (biota/evolutionary). Santos and Capellari's (2009) discussion on "reciprocal illumination" and "consilience" may guide us when moving from one quadrant to the other. Hennig's (1966) reciprocal illumination, in the context of phylogenetic systematics, implies that a particular relationship, based on some kind of evidence, may be tested ("illuminated") by comparison with another kind of evidence. Following this principle, it would be possible to bring together evolutionary and ecological analyses and examining whether deductions from them do agree or not. If discrepancies are evident, we may reexamine them and look for misinterpretations, mistakes, or missing information. The term consilience was coined by Whewell (1847) for situations where a theory proposed to explain a particular set of phenomena is found to provide a successful explanation for other phenomena, not considered during the construction of such theory. Thus, the robustness of a biogeographic theory may be evaluated by its possibilities of explaining new phenomena. Taxonomic and biotic analyses may benefit from their interaction while examining and contrasting the deductions of their particular hypotheses.

Lieberman (2003) challenged the traditional view of evolutionary and ecological biogeography as distinct and separate subdisciplines. He considered that there might be cases when evolutionary and ecological patterns coincide, for example, when different species have overlapping distributional areas. If these species show similar patterns of differentiation, it could be deduced that geological and climatic factors are playing an important role. And if, they show additionally coevolutionary relationships, there could be concordant evolutionary and ecological patterns.

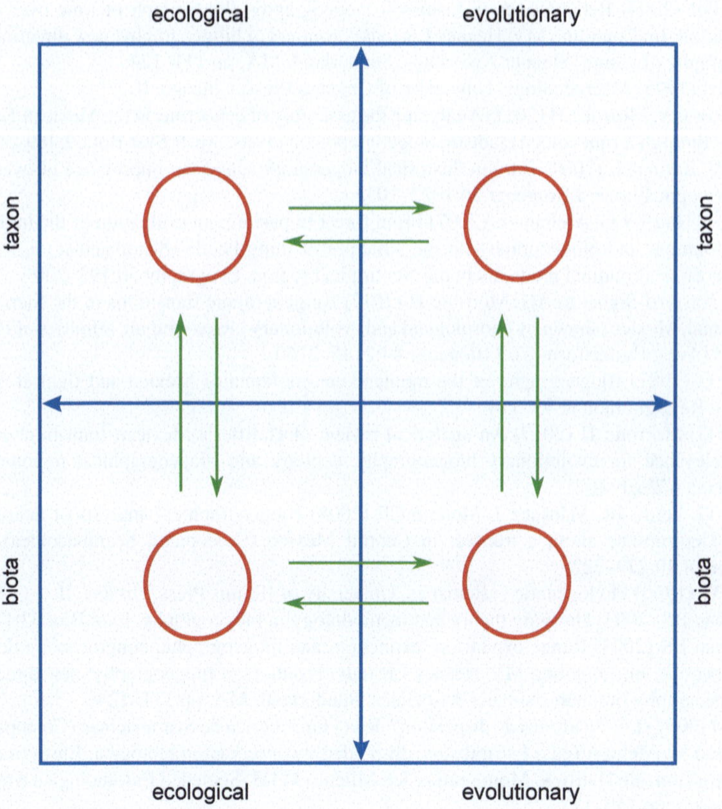

Fig. 6.1 Ecological/evolutionary biogeography and taxon/biota as the main axes delimiting quadrants for integrative biogeography

Weeks et al. (2016) proposed a protocol to analyze biotic assembly, integrating phylogenetic systematics and ecology. This protocol may be applied to transition zones, provided that their particularities are taken into consideration, and even extended to other systems, as mountains, islands, and habitat patches, among others. The reconstruction of cenocrons and horobiotas in a transition zone is a particularly complex issue involving phylogenetic, distributional, molecular, and geological studies. I would like to end this book with a challenge for the future: the development of a truly integrative biogeography, beyond the boundaries of evolutionary and ecological biogeography.

References

Aguado-Bautista O, Escalante T (2015) Cambios en los patrones de endemismo de los mamíferos terrestres de México por el calentamiento grlobal. Rev Mex Biodivers 86:99–110

Brooks DR (2004) Reticulations in historical biogeography: the triumph of time over space in evolution. In: Lomolino MV, Heaney LR (eds) Frontiers of biogeography: new directions in the geography of nature. Sinauer Associates, Sunderland, MA, pp 111–124
Brown JH (1995) Macroecology. University of Chicago Press, Chicago, IL
Corral-Rosas V, Morrone JJ (2017) Analyzing the assembly of cenocrons in the Mexican Transition Zone through a time-sliced cladistic biogeographic analysis. Austr Syst Bot 29:489–501
Crisci JV, Katinas L (2009) Darwin, historical biogeography, and the importance of overcoming binary oppositions. J Biogeogr 36:1027–1032
Escobar F, Halffter G, Arellano L (2007) From forest to pasture: an evaluation of the influence of environment and biogeography on the structure of dung beetle (Scarabaeinae) assemblages along three altitudinal gradients in the Neotropical region. Ecography 30:193–208
Ferro I, Navarro-Sigüenza AG, Morrone JJ (2017) Biogeographic transitions in the Sierra Madre Oriental, Mexico, shown by chorological and evolutionary biogeographic affinities of Passerine birds (Aves: Passeriformes). J Biogeogr 44:2145–2160
Halffter G (1987) Biogeography of the montane entomofauna of Mexico and Central America. Annu Rev Entomol 32:95–114
Halffter G, Morrone JJ (2017) An analytical review of Halffter's Mexican transition zone, and its relevance for evolutionary biogeography, ecology and biogeographical regionalization. Zootaxa 4226:1–46
Halffter G, Verdú JR, Márquez J, Moreno CE (2008) Biogeographical analysis of Scarabaeinae and Geotrupinae along a transect in Central Mexico (Coleoptera, Scarabaeoidea). Fragm Entomol 40:273–322
Hennig W (1966) Phylogenetic systematics. University of Illinois Press, Urbana, IL
Lieberman BS (2003) Unifying theory and methodology in biogeography. Evol Biol 33:1–25
Lieberman BS (2004) Range expansion, extinction, and biogeographic congruence: a deep time perspective. In: Lomolino MV, Heaney LR (eds) Frontiers of biogeography: new directions in the geography of nature. Sinauer Associates, Sunderland, MA, pp 111–124
Lobo JM (2007) Los "patrones de dispersión" de la fauna ibérica de Scarabaeinae (Coleoptera). In: Zunino M, Melic A (eds) Escarabajos, diversidad y conservación biológica: Ensayos en homenaje a Gonzalo Halffter,. Monografías 3er. Milenio M3M. Sociedad Entomológica Aragonesa, Zaragoza, pp 159–177
Lobo JM, Halffter G (2000) Biogeographical and ecological factors affecting the altitudinal variation of mountainous communities of coprophagous beetles (Coleoptera: Scarabaeoidea): a comparative study. Ann Entomol Soc Am 93:115–126
Moctezuma V, Halffter G, Escobar F (2016) Response of copronecrophagous beetle communities to habitat disturbance in two mountains of the Mexican Transition Zone: influence of historical and ecological factors. J Inst Conserv 20:945–956
Morrone JJ (2003) Las ideas biogeográficas de Osvaldo Reig y el desarrollo del "dispersalismo" en América Latina. In: Morrone JJ, Llorente J (eds) Una perspectiva latinoamericana de la biogeografía. Las Prensas de Ciencias, UNAM, Mexico City, pp 69–74
Morrone JJ (2009) Evolutionary biogeography: an integrative approach with case studies. Columbia University Press, New York
Morrone JJ (2011) La teoría biogeográfica de Florentino Ameghino y el carácter episódico de la evolución geobiótica de los mamíferos terrestres de América del Sur. Asoc Paleontol Argent Publ Esp 12:81–89
Morrone JJ (2015a) Halffter's Mexican transition zone (1962–2014), cenocrons and evolutionary biogeography. J Zool Syst Evol Res 53:249–257
Morrone JJ (2015b) Biogeographical regionalisation of the world: a reappraisal. Austr Syst Bot 28:81–90
Reig OA (1962) Las integraciones cenogenéticas en el desarrollo de la fauna de vertebrados tetrápodos de América del Sur. Ameghiniana 2:131–140
Reig OA (1981) Teoría del origen y desarrollo de la fauna de mamíferos de América del Sur. Museo Municipal de Ciencias Naturales Lorenzo Scaglia, Mar del Plata

Reyes-Castillo P (2003) Las ideas biogeográficas de Gonzalo Halffter: Importancia e impacto. In: Morrone JJ, Llorente Bousquets J (eds) Una perspectiva latinoamericana de la biogeografía. Las Prensas de Ciencias, UNAM, Mexico City, pp 99–108

Riddle BR, Dawson MN, Hadly EA, Hafner DJ, Hickerson MJ, Mantooth SJ, Yoder AD (2008) The role of molecular genetics in sculpting the future of integrative biogeography. Prog Phys Geogr 32:173–202

Ruggiero A, Ezcurra C (2003) Regiones y transiciones biogeográficas: Complementariedad de los análisis en biogeografía histórica y ecológica. In: Morrone JJ, Llorente-Bousquets J (eds) Una perspectiva latinoamericana de la biogeografía. UNAM, Mexico City, pp 141–154

Sanmartín I, Ronquist F (2004) Southern hemisphere biogeography inferred by event-based models: plant versus animal patterns. Syst Biol 53:216–243

Santos CMD, Capellari RS (2009) On reciprocal illumination and consilience in biogeography. Evol Biol 36:407–415

Savage JM (1966) The origins and history of the Central American herpetofauna. Copeia 1966:719–766

Savage JM (1982) The enigma of the Central American herpetofauna: dispersals or vicariance? Ann Missouri Bot Gard 69:464–547

Weeks BC, Claramunt S, Cracraft J (2016) Integrating systematics and biogeography to disentangle the roles of history and ecology in biotic assembly. J Biogeogr 43:1546–1559

Whewell W (1847) The philosophy of the inductive sciences founded on their history, vol 2, 2nd edn. John W. Parker, London

The manufacturer's authorised representative in the EU is Springer
Nature Customer Service Centre GmbH, Europaplatz 3, 69115 Heidelberg,
Germany. If you have any concerns regarding our products, please
contact ProductSafety@springernature.com

Printed and bound by CPI Group (UK) Ltd, Croydon, CR0 4YY
30/09/2026
02225689-0006